LESSONS LEARNED FROM HISTORY:

IMPLICATIONS FOR HOMELAND DEFENSE

by

Martha K. Jordan, Lt Col, USAF

Fredonia Books

Amsterdam, The Netherlands

Lessons Learned from History:
Implications for Homeland Defense

by
Martha K. Jordan

ISBN: 1-4101-0022-7

Copyright © 2002 by Fredonia Books

Reprinted from the original edition

Fredonia Books
Amsterdam, The Netherlands
http://www.fredoniabooks.com

All rights reserved, including the right to reproduce
this book, or portions thereof, in any form.

Disclaimer

The views expressed in this academic research paper are those of the author(s) and do not reflect the official policy or position of the US government or the Department of Defense. In accordance with Air Force Instruction 51-303, it is not copyrighted, but is the property of the United States government.

Contents

Page

DISCLAIMER II

PREFACE VII

ABSTRACT VIII

INTRODUCTION 1

AMERICAN HOMELAND DEFENSE: THE COLONIAL ERA THROUGH
THE POST-CIVIL WAR ERA..7
   Introduction 7
   The Colonial Era..8
      Colonial Militias 8
   The Revolutionary War 9
      Background to the Revolutionary War 9
      Militias in the Revolutionary War 10
      The First Continental Army..11
      Revolutionary War Lessons Learned..12
   The Interwar Years 1783-1812 11
      Interwar Militias 13
      Interwar Defense Strategy 13
   The War of 1812 15
      Background...15
      American Strategic Weaknesses in the War of 1812..16
      Land Campaigns...17
      Naval Campaigns..18
      War of 1812 Lessons Learned 19
   The Interwar Years 1815-1845 20
      Introduction 20
      The Coastal Defense Plan...21
      The Army in the Interwar Years...22
      Professional Officer Development 22
      Lessons Learned from the Interwar Years...23
   The Mexican War 24
      Background...24
      The Mexican War 24
      Strategic Problems in Aftermath of the Mexican War 25
   The Civil War 26
      Secession 26
      Union Strategic Advantages and Disadvantages 26
      Confederacy Strategic Advantages and Disadvantages.......................................27
      Scott's "Anaconda Plan" 28
      Lee's Strategic Mistake: Gettysburg 29
      Grant's strategic Victory: Vicksburg 30

  Grant's Final Plan 30
  Civil War Lessons Learned 31
 Post-Civil War 32
  Post-Civil War Defense Policy 32
  Origins of the National Guard 32
  Post-Civil War Threat Environment 33
 Analysis of American Homeland Defense: The Colonial Era Through the
  Post Civil War Era 33

AMERICAN HOMELAND DEFENSE: THE EXPANSIONIST ERA
 THROUGH WORLD WAR II 41
 The Expansionist Era: 1865-1898 41
  Background 41
  Alfred Thayer Mahan and Control of the Sea 42
  The U.S. Navy in the Expansionist Era 43
  The U.S. Army in the Expansionist Era 43
 The Spanish-American War 43
  Background 43
  Strategic Weaknesses in the Spanish-American War 44
  Naval Strategy in the Spanish-American War 45
  Land Operations in the Spanish-American War 46
  Failure to Coordinate Land and Naval Operations 46
  Spanish-American War Lessons Learned 47
 The Interwar Years Through World War I: 1898-1918 47
  Strategic Requirements Versus Resources 47
  The Dick Act: Federalizing the Militia 48
  The Beginning of U.S. Strategic Defense Planning 48
  War Plans ORANGE and BLACK 49
  The Wilson Administration: Homeland Defense Policy 49
  National Defense Act of 1916 50
  America's New Homeland Defense Requirements 51
  America Enters World War I 51
  Homeland Defense Strategy and Resources in World War I 52
  Problems in Mobilizing Industry for Homeland Defense in World War I 52
  Interwar Years Through World War I: Lessons Learned 53
 Interwar Years: 1919-1941 55
  Retrenchment and Isolationism 55
  Erosion of Military Readiness 55
  The Army-Air Corps and Homeland Defense 57
  Billy Mitchell's Analysis of U.S. Strategic Vulnerability 57
  The Court-Martial of Billy Mitchell 58
  The Beginning of Strategic Airpower Doctrine 59
  Roosevelt and Homeland Defense Policy: Hemispheric Defense 59
  The RAINBOW Plans 60
  Prelude to World War II: Efforts to Increase Readiness 60
  The Lend-Lease Act: American Homeland Defense Begins Overseas 61
 World War II 61

  Background 61
  Pearl Harbor: Homeland Defense Failure 62
  The U.S. Navy: Key to Pacific Victory 63
  The Battle of the Atlantic 64
  Victory in the Atlantic 65
  Strategy of Annihilation 66
  The Atomic Bomb 67
 Analysis of American Homeland Defense: The Expansionist Era Through
 World War II 68

AMERICAN HOMELAND DEFENSE: THE COLD WAR 76
 Truman: Containment 76
  Cold War: A Revolution in Homeland Defense 76
  Collective Security 77
  National Security Act of 1947 78
  Military Professional Education 79
  The Truman Doctrine 79
  The Marshall Plan 80
  Military Readiness Declines 81
 Truman and Civil Defense 82
 NSC 68 83
 The Korean War 84
 Massive Retaliation and The New Look 85
  Eisenhower and Strategic Alliances 87
  The Eisenhower Doctrine 87
  More Military Reductions 88
 The New New Look 88
 Eisenhower and Civil Defense 90
 Kennedy and Johnson: The Risks of Unintended Consequences 91
  Flexible Response 91
  Kennedy's Nuclear Policy 92
  Kennedy and Civil Defense 92
  Kennedy and Conventional Forces Buildup 93
  Failure to Develop Policy forUse of Conventional Forces 93
  Kennedy's Vietnam Escalation 94
  Unintended Consequences of Vietnam Esclation 94
  Alliance for Progress 95
  The Bay of Pigs 95
  The Cuban Missile Crisis 96
  Consequences of Failure to Develop Clear Strategy 97
 Johnson and Nixon: The Downward Spiral 97
  McNamara and "Mutual Assured Destruction" 97
  The Soviets and "Mutual Assured Destruction" 99
  McNamara and Anti-Ballistic Missile Capability 101
  The Beginnings of SALT and the End of U.S. Nuclear Superiority 104
  The Effect of Vietnam on U.S. Homeland Defense 106
 Ford and Carter: Defense Nadir 108

- Schlesinger Tries to Undo the Damage 109
- Effects of Recession on Homeland Defense 110
- Carter and the B-1 111
- Effect of Conventional Force Cuts 111
- Soviet Military Buildup 112
- The Iranian Hostage Crisis 112
- The End of SALT II 112
- The Carter Doctrine 113
- Desert One: Nadir of U.S. Defense Credibility 113
- The Ford-Carter Era: Lessons Learned 114
- Reagan and Bush: Defense Resurgence 115
  - Defense Buildup and the End of Detente 115
  - Reagan's National Security Strategy 115
  - Conventional Forces Buildup 116
  - International Terrorism and Homeland Defense 118
  - Security Assistance and Homeland Defense 121
  - The Reagan Doctrine 122
  - Reagan and Arms Control 126
  - Bush and Arms Control 129
  - Military Force Reductions 131
  - The Powell Doctrine 132
- Analysis of American Homeland Defense: The Cold War 133

AMERICAN HOMELAND DEFENSE: THE CLINTON ERA 152
- Background 152
- Terrorism: the Greatest Challenge to U.S. Homeland Defense 153
- Ballistic Missile Proliferation 157
- Russia's Command and Control of Nuclear Weapons 160
- Clinton and National Missile Defense 161
- Defense Against Conventional Threats 162
- Domestic Improvements to Homeland Defense 163
- Analysis of American Homeland Defense: The Clinton Era 169

LESSONS LEARNED FROM AMERICAN HOMELAND DEFENSE HISTORY 179

CONCLUSION: AMERICAN HOMELAND DEFENSE IN THE 21ST CENTURY 189

GLOSSARY: 206

BIBLIOGRAPHY 212

Preface

The U.S. faces a host of new threats to our vital interests, including proliferation of cruise missiles, weapons of mass destruction, ballistic missile technology, questions about Russia's control of strategic nuclear forces, and an increasing tendency for terrorists to conduct mass casualty attacks. These new threats require a comprehensive, deterrence-oriented homeland defense program. Lessons learned from a study of U.S. homeland defense history will provide a basis for recommending improvements to homeland defense policy and strategy.

My year at the Center for Strategic and International Studies (CSIS) provided a superb opportunity to study homeland defense policy issues. CSIS sponsors a major program to study all facets of homeland defense policy and strategy. I hope this paper will provide a historical foundation for understanding the problems we currently face in developing a coherent homeland defense policy. This paper would not be possible without the superb guidance and support of CSIS Homeland Defense project members. I especially wish to thank Mr. Joe Collins, Mr. Dan Goure, and Dr. Kurt Campbell for their excellent suggestions and support. A special note of thanks goes to Dr. William Green for his expertise in Cold War strategy and policy, and to Col Robert Sutton, Air University, for his excellent advice on current U.S. homeland defense efforts and counter proliferation.

Abstract

Homeland defense is a basic function of our government, and has been since the colonial era. Over 100 years before our Constitution established as a fundamental responsibility of government to "provide for the common defense," American colonial leaders recognized the need for a homeland defense capability and established colonial militias to fight both internal and external threats. As the U.S. grew in economic and political power, our homeland defense needs changed, as well. By the 20th century, homeland defense encompassed not only defense of U.S. territory and population, but overseas possessions, access to critical lines of communication, natural resources, and trade. The Cold War again qualitatively and quantitatively changed the concept of homeland defense. The U.S. had to defend itself against the Soviet Union's global hegemonic ambitions and growing nuclear arsenal, so homeland defense required much broader measures than ever before. This paper discusses the development of U.S. homeland defense from the colonial era to the present and analyzes homeland defense policy failures and successes. Analysis shows common problems with homeland defense policy: failure to develop an overarching, viable homeland defense strategy, failure to provide appropriate military resources to meet strategic requirements, and failure to understand and prepare for emerging threats. Analysis is used as a basis for suggesting improvements to homeland defense capability in the 21st century.

Chapter 1

Introduction

If we judge the future from the past, we perceive that though there may be no war, we must not rest at ease.

—Admiral Marquis Togo Heihachiro

Homeland defense is an issue that recently received renewed, urgent interest due to the spread of international terrorism, advanced weaponry, and weapons of mass destruction among nations and groups hostile to the United States and its allies. The United States' overwhelming political, military, and economic power, coupled with its fortunate geographic position between two oceans to the east and west, and friendly neighbors to the north and south, are not an adequate defense against these burgeoning threats. In fact, former President Clinton stated that "within the next ten years, there [is] a 100 percent chance of a chemical or biological attack in our country."[1]

Developing an appropriate policy and strategy to protect the U.S. against these new threats is hampered by the fact that there is as yet[2] no nationally accepted definition of homeland defense. The National War College notes that the term has been used

---

[1] Quoted in Joseph J. Collins and Michael Horowitz, *Homeland Defense: A Strategic Approach* (Washington, D.C.: Center for Strategic and International Studies, 2000), 2; on-line, Internet, 9 March 2001, available from http://www.csis.org/homeland/reports/hdstrategicappro.pdf.

[2] Joseph J. Collins and Michael Horowitz, *Homeland Defense: A Strategic Approach* (Washington, D.C.: Center for Strategic and International Studies, 2000), 42; on-line, Internet, 9 March 2001, available from http://www.csis.org/homeland/reports/hdstrategicappro.pdf.

"interchangeably with national missile defense, counter-terrorism, "consequence management" (see Glossary for definition) or the after-effects of the use of a weapon of mass destruction, military support to civil authorities and information warfare."[3] The U.S. Army Training and Doctrine Command defines homeland defense as: "…protecting our territory, population and critical infrastructure at home by deterring and defending against foreign and domestic threats; supporting civil authorities for crisis and consequence management; helping to ensure the availability, integrity, survivability, and adequacy of critical national assets."[4] The Center for Strategic and International Studies Working Group defines homeland defense as "…the defense of the United States' territory, critical infrastructure, and population from direct attack by terrorists or foreign enemies operating on our soil…"[5]

The problem with defining homeland defense lies in its boundaries. Is homeland defense strictly confined to domestic measures to deter and, if necessary, respond to attacks on American soil, or is it a much broader concept, essentially a synonym for "national security?"[6] The Department of Defense Dictionary of Military Terms defines "national security" as: "A collective term encompassing both national defense and foreign relations of the United States. Specifically, the condition provided by: a. a military or defense advantage over any foreign nation or group of nations, or b. a favorable foreign relations position, or c. *a defense posture capable of successfully*

---

[3] National War College, NWC Course 5605, Military Strategy and Operations, "Topic 12: Homeland Defense," 24 March 2000, n.p.; on-line, Internet, 29 August 2000, available from http://www.ndu.edu/ndu/nwc/5605SYL/Topic12.html.

[4] U.S. Army Training and Doctrine Command (TRADOC), *White Paper: Supporting Homeland Defense*, 18 May 1999, 1-2; on-line, Internet, 1 March 2001, available from http://www.fas.org/spp/starwars/program/homeland/final-white-paper.htm.

resisting hostile or destructive action from within or without, overt or covert [Italics added].

Review of the definitions of "homeland defense" and "national security" provide much latitude for interpretation regarding the limits of homeland defense. One could effectively argue that there is no essential difference between "national security" and "homeland defense." Or, if one wishes to adopt a narrower view of homeland defense, then it could be viewed as strictly domestic measures to protect the U.S. against attack. But taking such a narrow view defeats the whole purpose of homeland defense.

If homeland defense encompasses protecting the U.S. against the plethora of threats to our territory, critical infrastructure, and population, then effective policy and strategy for homeland defense *must* use the entire range of instruments of U.S. national power—economic, political, informational, diplomatic, and military—both domestically *and* internationally. Why? Because the most dangerous, urgent threats to the U.S. don't necessarily originate within our borders. International terrorism (in all of its forms, such as bombings, shootings, cyber attacks, kidnappings, and insurgencies against strategically vital allies), proliferation of ballistic missiles and weapons of mass destruction, and problems in control and accountability for nuclear materials most often originate in foreign countries (with some exceptions). Therefore, a viable homeland defense strategy and policy *must* emphasize the use of U.S. power to degrade and destroy these threats before they reach the U.S. homeland. If we have to resort to domestic response after an

---

[5] Frank Cilluffo et al., *Defending America in the 21st Century*, (Washington, D.C.: Center for Strategic and International Studies, 2000) 2; on-line, Internet, 9 March 2001, available from http://www.csis.org/homeland/reports/defendamer21stexesumm.pdf.
[6] National War College, "Topic 12: Homeland Defense," n.p.

attack, then our homeland defense strategy has already failed. In essence, homeland defense strategy is now synonymous with national security strategy.

An historic study of U.S. homeland defense since the American colonial era provides a basis for understanding both the domestic and international dimensions of U.S. homeland defense, and why it became essentially synonymous with "national security." Furthermore, an historical study of U.S. homeland defense policy and strategy offers some insights into modern homeland defense policy and strategy problems; namely, that our current problems in developing effective homeland defense policy had their origins over 300 years ago when America was still a collection of colonies.

Over 100 years before the U.S. Constitution established as a fundamental responsibility of government to "provide for the common defense,"[7] American colonial leaders recognized the need for a homeland defense capability and established colonial militias to fight both internal and external threats to the colonists. Homeland defense needs at that time were much more narrow in scope than today. Protection against raids on the colonies and encroachment by foreign powers comprised the scope of colonial homeland defense.

As our nation grew in economic and political power, the scope of homeland defense changed. The welfare and safety of the U.S. population, protection of our expanding territory on the continent and overseas possessions, and safe, unfettered access to vital resources and trade routes all became critical to the economic and political survival of our nation. Homeland defense thus became more complicated to cope with these new

---

[7] National Archives and Records Administration, *Constitution of the United States of America*; on-line, Internet, 17 February 2001, available from http://www.nara.gov/exhall/charters/constitution/constitution.html.

requirements. Defense of overseas possessions, protection of critical trade routes and sea lines of communication all became an essential part of homeland defense.

But our tradition as independent people and as a democratic nation born of a revolution against the injustices and excesses of a monarchical global military power shaped and often limited homeland defense policies and resources. The government tried to "provide for the common defense" without placing an undue burden on the people or betraying the values on which our country was founded. As a result, our government was traditionally reluctant to fund a strong homeland defense capability until faced with a national crisis.

Other factors shaped historic homeland defense policies. A fortunate geographic position placed oceans to our east and west and relatively peaceful neighbors to the north and south. In addition, our people have a historic suspicion of large standing armies, and, except in times of national crisis, have not supported large monetary defense expenditures.[8] These policies and beliefs helped maintain our democratic nation for over 200 years.

However, as history shows, these policies were not without risk. For example, failure to prepare an adequate homeland defense nearly cost us our nation during the War of 1812. Later conflicts, both inside and outside U.S. borders, highlighted the importance of maintaining a well-trained, well-equipped regular force. But it was not until after World War II, with the beginning of the Cold War, the introduction of nuclear weapons, and the emergence of America as a truly global power, that America developed a large, permanent, globally capable military force, and entered into long-term formal military alliances with non-communist allies. Our homeland defense policy changed to meet the challenges of a significantly changed geopolitical environment.[9]

Since the end of the Cold War, our homeland defense requirements changed, again. The oceans, our friendly neighbors, strategic alliances, atomic weapons, and large military force are no longer an effective deterrent to new threats to our nation. Terrorism, ballistic missiles, and weapons of mass destruction are new threats that cannot be defeated using the previous centuries' strategies. As U.S. Army historians suggested, "The increasing complications of modern warfare, the great rapidity with which attacks can be launched with modern weapons, and the extensive overseas commitments of the United States have negated the traditional American habit of preparing for wars after they have begun."[10]

The myriad new international threats to the U.S. homeland and our vital interests demands that the U.S. must now approach homeland defense as both a domestic and international issue—essentially, it is now synonymous with national security.

Analysis of the historic successes and failures of U.S. homeland defense policies from the colonial era to the present suggest a basis for an effective homeland defense. What is needed is an overarching national strategy, encompassing both domestic and international policies and plans, to meet the 21st century needs of U.S. homeland defense.

---

[8] Office of the Chief of Military History, United States Army, "Chapter1, Introduction," in *American Military History* (Washington, D.C.: Office of the Chief of Military History, 1988), 3-4, 14-15; on-line, Internet, 30 November 2000, available from http://www.army.mil/cmh-pg/books/amh/amh-01.htm
[9] Office of the Chief of Military History, "Chapter 1, Introduction," 15-16.
[10] Office of the Chief of Military History, "Chapter 1, Introduction," 17.

Chapter 2

American Homeland Defense: The Colonial Era Through The Post-Civil War Era

Men dragged from the tender Scenes of domestick life; unaccustomed to the din of Arms; totally unacquainted with every kind of Military skill, which being followed by a want of confidence in themselves, when opposed to Troops regularly train'd, disciplined, and appointed, superior in knowledge, and superior in Arms, makes them timid, and ready to fly from their own shadows.

—George Washington

Introduction

From the colonial era into the 19th century, our country had only a very limited homeland defense capability. We did not have the means to exploit natural resources, our population was limited (about 3 ½ million when the Constitution was drafted), and communications and manufacture were primitive. Hence, our homeland defense objectives were limited, as well. Internal defense against raids on towns and settlements, and defense against encroachment of other colonial powers were the limits of homeland defense objectives.[11]

Our limited defense requirements, coupled with popular suspicion of large standing armies and insistence on local control over militias, resulted in a basically ad hoc approach to homeland defense. Although this approach saved money and resources for other priorities, in our new nation's struggle to increase its economic and political power,

---

[11] Charles J. Hitch, *Decision-Making for Defense* (Berkeley, California: University of California Press, 1965), 5.

ad hoc homeland defense policies nearly cost the nation its independence in both the Revolutionary War and the War of 1812, and ensured a long, drawn-out bloodbath in the Civil War.

The Colonial Era

From the earliest days of the North American colonies, warfare between Native Americans and European colonizers, and between European governments over colonial claims, was a not infrequent fact of life. Incompatibility of cultures between Native Americans and European settlers, invasion of Native American territory, and competition for land and commercial resources all contributed to frequent, and in some places, incessant conflicts.[12] However, the European powers' desires to establish permanent claim to lands and colonies in North America did not initially translate to a desire to expend large sums of money for their defense. Civil and dynastic wars on the European continent (such as the English Civil Wars (1642–1651) and the War of the Grand Alliance (1688-1697)) drained royal treasuries, and the enormous expense of maintaining standing armies in the colonies precluded long-term commitments of armies to protect the colonies.[13] Hence, the early colonies bore much of the responsibility and expense of defending themselves.

Colonial Militias

To defend themselves against internal and external threats, colonists developed a militia similar to the Europeans. By 1638, British colonists took the English model of a "citizen soldier" and used it to develop their own militias. In 1638, the first American military unit, The Ancient and Honorable Artillery Company, was born in Boston. In the

American tradition of wariness toward standing armies, some colonists expressed suspicion concerning this new unit's desire to take political power.[14] Massachusetts was also the first to initiate compulsory military training for males in 1643, and other British colonies (except Pennsylvania, which had a significant Quaker influence) soon did the same.[15]

Militia members had to provide their own arms, and were usually neither well trained, nor well equipped, nor well disciplined, compared to European professional armies.[16] Additionally, the colonies did not coordinate organization or training of their militias. Each colony had a separate militia that protected its own interests; cooperation only occurred when two or more colonies had a shared interest in defeating a threat.[17] Generally, those militias on the frontier, which were in frequent conflict with Native American tribes, were more proficient out of necessity.[18]

The skirmishes and wars (such and the French and Indian War) between rival colonial powers, and between colonists and Native Americans, set the stage for American defense policy and practice through the 19th century. Suspicious of standing regular armies, Americans relied on militias for defense and volunteers for special, large-scale expeditions. Furthermore, with the strong individualism of each colony, each organized and trained its own militia forces. Civilians maintained strict control over the militias.

---

[12] R. Ernest Dupuy and Trevor N. Dupuy, *The Harper Encyclopedia of Military History from 3500 B.C. to the Present*, 4th ed. (New York: HarperCollins Publishers, 1993), 658.
[13] Dupuy and Dupuy, 597-598, 602-606, 658-659.
[14] Dupuy and Dupuy, 660.
[15] Dupuy and Dupuy, 658-659.
[16] Office of the Chief of Military History, United States Army, "Chapter 2, The Beginnings" in *American Military History* (Washington, D.C.: Office of the Chief of Military History, 1988), 28-30; on-line, Internet, 30 November 2000, available from http://www.army.mil/cmh-pg/books/amh/amh-02.htm
[17] Office of the Chief of Military History, "Chapter 2, The Beginnings," 28-29.
[18] Office of the Chief of Military History, "Chapter 2, The Beginnings", 30.

These policies set the stage for the outcome of major battles with British regular forces during the Revolutionary War.[19]

The Revolutionary War

Background to the Revolutionary War

As the colonies grew in population and wealth, so, too did separatist sentiments. These sentiments turned to action when the British government levied extra taxes on the colonies to pay the costs of maintaining British soldiers in the colonies. Colonists' anger with the increased tax burden culminated in the December 1773 Boston Tea Party. In response, the British closed Boston's port and put Massachusetts under military rule.[20]

This was a massive strategic blunder, for these actions galvanized the colonies to act together to coerce the British government into repealing the port closure and military rule in Massachusetts. The colonies soon formed revolutionary committees in nearly every county and town. These committees soon became *de facto* local governments—they also took control of the colonial militias, from which sprang the fighting forces of the American Revolution.[21]

Militias in the Revolutionary War

Colonial era military policies left the colonies ill prepared to fight and win against British military forces. Coordination of training and tactics between colonial militias was nonexistent. Militias trained infrequently, and lacked the discipline of British forces.[22]

---

[19] Office of the Chief of Military History, "Chapter 2, The Beginnings", 39-40.
[20] Office of the Chief of Military History, "Chapter 3, The American Revolution: First Phase," in *American Military History* (Washington, D.C.: Office of the Chief of Military History, 1988), 41-42; on-line, Internet, 30 November 2000, available from http://www.army.mil/cmh-pg/books/amh/amh-03.htm.
[21] Office of the Chief of Military History, "Chapter 3, The American Revolution: First Phase," 42.
[22] Office of the Chief of Military History, "Chapter 2, The Beginnings," 39-40.

Furthermore, the colonies had no prepared strategy to defeat British forces. Instead, events drove strategy.[23]

The precipitating event of the Revolutionary War occurred at Lexington on 18 April 1775. The British commander in Boston, General Gage, learned of a collection of military supplies secretly stored at Concord. He sent troops from Boston to destroy the stores, but halfway to Concord, the British were met by a group of colonial militia. The ensuing firefight left 8 militia dead and 10 wounded. Word of the fight spread quickly, and the British were continually ambushed by militia on their way back from Concord to Boston, resulting in over 200 British casualties. This one event sparked armed rebellion throughout the colonies. Militia from other New England colonies quickly came to the aid of the Massachusetts militia, and forces from Vermont and Connecticut captured key British forts between New York and Canada.[24]

The First Continental Army

The speed of the developing revolution forced the Second Continental Congress to plan and organize a war using all colonial resources.[25] The Continental Congress started organizing for war, and on 14 June 1775, the Continental Congress created the first Continental Army and named George Washington Commander in Chief.[26] This set a pattern for American homeland defense for many years to come—hastily assembling a group of citizen soldiers into an army after a crisis has already begun.

Although Washington knew his forces were significantly weaker than the British, he at first insisted on using conventional tactics and strategy, ignoring the advice of some

---

[23] Russell F. Weigley, *The American Way of War* (Bloomington, Indiana: Indiana University Press, 1973), 7.
[24] Office of the Chief of Military History, "Chapter 3: The American Revolution: First Phase," 42-43.
[25] Office of the Chief of Military History, "Chapter 3, The American Revolution: First Phase," 42-43.

senior officers who advised him to conduct a war of attrition (in modern terms, a "guerrilla" war).[27] Of course, his soldiers were neither well trained nor disciplined enough to execute strategy and tactics of European-style conventional warfare (disciplined troops maneuvering in formation on a battlefield). After his defeat in the defense of New York City in 1776 (and later at the Battle of Brandywine in 1777), he realized he could not hope to win by facing the main body of British forces through conventional style warfare.[28] He tried to avoid confrontations with massed British forces, and relied instead using hit and run tactics to wear down the British Army and undermine British popular support for the war.[29] When his army did attack the British, he attacked only portions of the British forces. After New York, he used a defensive strategy aimed at preserving his army.[30]

The 1778 alliance with the French finally gave Washington the opportunity to bring an end to the war. The assistance of the French Navy diluted British strength, since the British not only had to fight the French in the colonies, but also had to maintain part of their fleet at home to prevent the French from attacking England.[31] The end came at Yorktown, when the Americans and French trapped Cornwallis' forces on land, and French naval forces prevented his escape by sea.[32]

Revolutionary War Lessons Learned

The Revolutionary War provided some important lessons in homeland defense for the new country. First, a central government was needed to harness and coordinate the

---

[26] Office of the Chief of Military History, "Chapter 3, The American Revolution: First Phase," 46-47.
[27] Peter Paret, ed., *Makers of Modern Strategy from Machievelli to the Nuclear Age* (Princeton, N.J.: Princeton University Press, 1986), 410-411.
[28] Paret, 411-412.
[29] Weigley, 5.
[30] Weigley, 11-12.
[31] Weigley, 38-39.

resources of the colonies for a major war. The new Constitution reflected this fact, giving Congress the power to call up the militia, "raise armies and navies", and collect taxes to pay for these forces.[33] Additionally, Washington recognized the importance of maintaining a consistently organized and trained militia to be called to service in a crisis.[34] In 1784, Washington suggested funding a small regular army, backed up by a well-trained militia.[35]

The Interwar Years: 1783-1812

The Revolutionary War had exhausted our young nation's military capability, and the government did not have the funds to pay for a standing military. Furthermore, fear of a standing army (and navy) was prevalent throughout the states. The Continental Navy no longer existed, and the Army was soon reduced to a single regiment. Had an enemy mounted a major attack on either the east coast or from the Mississippi River, the young government would have had difficulty in successfully defending against such an attack.[36] Fortunately for the U.S., no serious threat to the new nation surfaced during the 1780's, so a stronger military force was not required, and the national defense strategy remained passive.[37]

Interwar Militias

The drafters of the U.S. Constitution had a deep mistrust of large regular armies, thanks to their experience with the British prior to and during our Revolutionary War.

---

[32] Dupuy and Dupuy, 787-788.
[33] Office of the Chief of Military History, "Chapter 4, The Winning of Independence, 1777-1783," in *American Military History* (Washington, D.C.: Office of the Chief of Military History, 1988), 99-100; on-line, Internet, 30 November 2000, available from http://www.army.mil/cmh-pg/books/amh/amh-04.htm
[34] Office of the Chief of Military History, "Chapter 4, The Winning of Independence, 1777-1783," 100.
[35] Paret, 412.
[36] Weigley, 41-42.
[37] Weigley, 41.

The anti-federalists especially were opposed to a standing army, believing that large armies could be used as a means of internal repression.[38] Hence, protection of U.S. territory from foreign aggression and internal rebellion became the responsibility of state militias. The Constitution gave Congress the authority to call up the militia in the event of invasion.[39]

According to the Militia Act of 1792, all adult male citizens were automatically members, had to be armed, and were subject to drill.[40] However, the federal government provided no funding for state militias, so the state militias continually suffered from lack of funding, equipment, manpower, and inadequate training; hence, they were not effective deterrents to a potential enemy attack on the U.S. homeland.[41]

Interwar Defense Strategy

The French Revolution, and the resulting hostilities between Great Britain and France, dragged the U.S. into the conflict and necessitated a more aggressive homeland defense strategy. Both Great Britain and France started interfering with U.S. shipping, and when North African states increased piracy against U.S. ships, Congress took action to improve homeland defense. In 1794, Congress ordered restoration of Revolutionary War coastal forts and construction of 16 new forts to protect vital ports and U.S. Navy vessels. The commissioning of new frigates marked a new, more aggressive policy in homeland defense. The new defense strategy was designed to protect our most vital ports

---

[38] Chuck Dougherty, "The Minutemen, The National Guard, and the Private Militia Movement: Will the Real Militia Please Stand Up?" *John Marshall Law Review*, Summer 1995, pp. 959-968. Available from http://www.saf.org/LawReviews/Dougherty1.html.
[39] Dougherty, pp. 965-966.
[40] U.S. Congress. *The Militia Act of 1792*, 2nd Cong., 1st sess., 2 May 1792. On-line, Internet, 13 March 2001, available from http://www.constitution.org/mil/mil_act_1792.htm.
[41] Dougherty, 969.

and harbors with fixed fortifications, while using warships to protect merchant shipping and stop an attacking fleet from reaching our harbors.[42]

By 1794, Congress authorized creation of the Navy Department, and in 1798, President John Adams appointed Benjamin Stoddert to oversee the building of the U.S. Navy and development of a comprehensive strategy based on naval power projection.[43] Stoddert recommended building 12 74-gun ships-of-the-line and 24 frigates (frigates were smaller warships with about 24-44 guns). Although this small Navy could not compete offensively with either Great Britain's or France's fleets of warships, Stoddert figured that it would take more that twice that many warships for an invading force to successfully land in America. Additionally, he felt that the distance between America and Europe gave us a strategic defensive advantage.[44]

Although Congress approved money to build 6 of the 12 proposed 74-gun ships, the program ended when France and the U.S. agreed to peace terms. The new President, Thomas Jefferson, did not want a large standing military force and wanted to reduce defense expenditures. Instead, he favored a fleet of small drought gunboats for harbor defense. He felt these would afford flexibility by enabling quick movement from port to port, as necessary. Furthermore, in time of war, these craft, simple to operate, could be manned by citizen militia, which he strongly preferred over maintaining a large standing military. Eventually, Congress appropriated enough money to build 167 of these craft.[45]

---

[42] Weigley, 41-43.
[43] Weigley, 43.
[44] Weigley, 44-45.
[45] Weigley, 45.

The Revolutionary War had proven the importance of the Navy; protection of our coastlines and overseas commerce were key to American security and prosperity.[46] But Jefferson's defense policies left America unable to easily defend itself against any major maritime threat. Instead of building a strong Navy, Congress had poured money into strengthening harbor fortifications. As the War of 1812 loomed in the foreground, our Navy consisted of but 16 ships (6 of them frigates) in addition to the 167 gunboats. Furthermore, there was no agreed-upon strategy for their employment during war.[47]

The Army's situation was no better. Although Congress had voted to increase the size of the Army just before the start of the War of 1812, years of neglect had left the Army unprepared for a major war. Thus, as America was about to face a war with a major maritime power, our homeland defenses consisted of a series of fixed fortifications, an Army of ill-equipped, untrained soldiers, and a fleet of a few frigates and some small gunships. Just as in the Revolutionary War, America's limited economic resources, coupled with the government's reluctance to maintain a large standing military force, led to passive defense policies which left the country ill-prepared for the coming war with Great Britain.[48]

The War of 1812

Background

The war began over British seizure of American ships and seamen, as well as our desire to expand our territory. England was at war with France again, and the British seized American merchant ships and cargoes for violating the British blockade of Europe. Furthermore, they impressed American seamen into the British navy on the pretext of

---

[46] Weigley, 40.

retrieving deserters from the British Navy. On the American frontier, the populace felt that the British were arming the Native Americans to prevent American expansion, and frontiersmen felt that driving the British out of Canada would solve the problem.[49]

Initially, Jefferson's administration responded to British provocations by initiating an embargo of American trade in 1807, which ruined many New England ship owners and caused an economic depression. Thus, while Americans on the western frontier supported going to war, New Englanders blamed Jefferson for their troubles.[50]

America entered the War of 1812 without a clear consensus of either the populace or the power to achieve its objectives. When President Madison asked Congress to declare war in June 1812, he wanted to force Great Britain to respect American neutrality and rights at sea. But former President Jefferson's military policies had left America without a navy capable of compelling the British to respect American rights at sea.[51]

American Strategic Weaknesses in the War of 1812

Instead, Congress authorized the invasion of Canada in hopes of eliminating the British from the North American continent, thereby enhancing homeland security.[52] But our Army was not up to the task, either. The Regular Army plus extra recruits numbered less than 17,000 men.[53] These were directed against about 7,000 British and Canadian forces stretched along the 900-mile Canadian frontier.[54]

---

[47] Weigley, 46.
[48] Weigley, 46.
[49] Office of the Chief of Military History, "Chapter 6, The War of 1812," in *American Military History* (Washington, D.C.: Office of the Chief of Military History, 1988), 122-124; on-line, Internet, 30 November 2000, available from http://www.army.mil/cmh-pg/books/amh/amh-06.htm
[50] Office of the Chief of Military History, "Chapter 6, The War of 1812," 122-124.
[51] Weigley, 46-47.
[52] Weigley, 46-47.
[53] Office of the Chief of Military History, "Chapter 6, The War of 1812," 125.
[54] Weigley, 47.

Another significant weakness in our defense capability was our transportation system. America did not have an adequate system of roads to transport supplies to all the different areas of fighting, so supplies were constantly short. Troops and their animals did not have enough blankets, ammunition, food, or shelter. Furthermore, no standardized system of supply existed, and district Quartermaster representatives acted independently; there was never any centralized control of critical supplies.[55]

As in the Revolutionary war, the political objective outstripped our military capability. Furthermore, no tenable overarching strategy to achieve American objectives existed, resulting in failures of several important land campaigns.

Land Campaigns

The first strategically significant failure was the unsuccessful invasion of Canada. Instead of using our military forces to attack one strategic point (Montreal) to cut off western Canada, the American forces attacked along the western Canadian frontier in a series of strategically useless, failed efforts that resulted in the Americans being driven from Canada before 1813.[56] The American failure to drive the British out of Canada in the initial campaign was accompanied by a series of similar disasters: capture of two key northern American forts, the surrender of Detroit, failed attacks at Queenston, Lake Champlain, and the massacre at Frenchtown.[57] The American recapture of Detroit and General William Henry Harrison's victory at the Battle of the Thames (which caused the collapse of the Great Britain's allies, the Indian confederacy) were successful, but the government failed to exploit these victories. Instead, the War Department ordered

---

[55] Office of the Chief of Military History, "Chapter 6, The War of 1812," 138-139.
[56] Weigley, 48.
[57] Dupuy and Dupuy, 870-873.

Harrison's militia disbanded, and the regulars were sent to another command. Harrison was so angry, he resigned.[58]

When the American forces finally attacked Montreal in the autumn of 1813, the expedition was another disaster. The two commanders, Hampton and Wilkinson, disliked and distrusted each other, failed to coordinate their efforts, and neither had enough resources to capture Montreal alone. Hampton's forces retreated after first contact with the British, and Wilkinson soon followed after a severe beating at the hands of British forces.[59] The Montreal expedition had pulled needed American forces from the Niagara area, and the British soon recaptured Fort George and took Fort Niagara from weakened American forces; the burning of Buffalo soon followed.[60]

Despite some successes in 1814, including Andrew Jackson's southern campaigns in Alabama and Pensacola, and Jacob Brown's and Winfield Scott's victories on the Niagara front at Chippewa and Lundy's Lane,[61] the land campaigns were generally not strategically successful. Not until Jackson's successful defense of New Orleans in 1814 did the land forces win a strategic victory, for this action prevented a British invasion from the south.

Naval Campaigns

The fledgling American Navy was not in a position to defeat the British in traditional engagements, either. In 1812, America's Navy consisted of 3 large 44-gun frigates, 3

---

[58] Dupuy and Dupuy, 870-873.
[59] Office of the Chief of Military History, "Chapter 6, The War of 1812," 134-135.
[60] Office of the Chief of Military History, "Chapter 6, The War of 1812," 134-135.
[61] Dupuy and Dupuy, 875-878.

small frigates, and 14 other vessels. In 1813, Congress had voted to build 10 more fighting ships, but this effort was too little, too late.[62]

The British had more than 600 fighting ships, most of which were supporting their war effort in Europe until 1814. Only 8 British fighting ships were in American waters initially, but by May 1813, the British moved from a defensive to offensive naval strategy in America, and had moved enough warships to the American war effort to blockade the Atlantic and Gulf coasts except New England (which did not support the war).[63] With the end of war in Europe in 1814, however, Great Britain turned its full attention to the war with America, and proclaimed the entire American coast under blockade.[64]

America's naval inferiority could not hope to compete with the British in traditional naval engagements; our Navy was forced to concentrate efforts on raiding British merchant vessels. Privateers were especially successful in this campaign.[65] Privateer attacks against British ships kept American ports open during the initial months of the war.[66] With the blockade in full force in 1814, privateers still were able to inflict substantial harm to British war efforts; they wreaked havoc on British commerce, capturing over 800 British merchant ships, and forcing much naval traffic on Great Britain's coasts to travel under convoy.[67] But despite the successes of the privateers, British naval forces controlled access to key American coastal areas, and by August 1814, the U.S. Treasury was nearly bankrupt, and the British entered Washington and burned the capitol.[68]

---

[62] Weigley, 53.
[63] Weigley, 49-51.
[64] Weigley, 51.
[65] Dupuy and Dupuy, 874.
[66] Office of the Chief of Military History, "Chapter 6, The War of 1812," 124-126.
[67] Dupuy and Dupuy, 878-879.
[68] Dupuy and Dupuy, 873-874, 876.

The Americans however, won a strategically significant victory at Lack Champlain in September 1814. The American naval contingent successfully protected the American position at Plattsburg, and forced the surrender of the British naval squadron, which had been acting in concert with British land forces in an attempt to take Plattsburg—a key corridor of invasion from the north.[69]

The British naval defeat at Lake Champlain was the most important engagement of the war, for it not only prevented a British conquest from the north, but it added impetus to peace negotiations to end the war. The British public was already weary of the war, and this defeat, following closely on the heels of the failed British assault on Baltimore, aided American negotiators in obtaining satisfactory peace terms.[70]

War of 1812 Lessons Learned

The War of 1812 provided several important lessons regarding homeland defense strategy. Most important, it illustrated the fact that American military commanders lacked sufficient education and experience to plan and execute a coordinated campaign strategy. As Secretary of War John Calhoun pointed out, war is an art, and America needed experienced military leaders with a comprehensive understanding of strategy.[71] America had entered the war with no systematic means of educating military leaders on strategy and doctrine. At the time, West Point was "a neglected foundling;" no higher schools of military education and training existed in America. Prior to the War of 1812, nothing had forced American leaders to become interested in education of a professional

---

[69] Dupuy and Dupuy, 879.
[70] Office of the Chief of Military History, "Chapter 6, The War of 1812," 143-144.
[71] Weigley, 55.

officer corps. The War of 1812 clearly illustrated the need for fundamental and advanced education in strategy and doctrine for its officer corps.[72]

The war also illustrated the need for militarily literate civilian leadership. Years of passive defense policies and Congressional refusal to fund an adequate standing army or navy left America unprepared to fight and win against a more formidable foe. Logistics and transportation within and between the states were inadequate to the task of provisioning an army for a protracted land campaign across a several hundred-mile front; the navy and army were far too small to be effective against so powerful an adversary as Great Britain; fixed fortifications did not prevent the enemy from blockading our coast.

The war also proved that a trained, experienced standing professional army was a necessity. Both militia and Regular forces suffered many defeats at the hand of superior British forces, thanks to lack of training, lack of enough manpower, lack of skilled leadership, and lack of a dependable logistics system.[73]

The Interwar Years: 1815-1845

Introduction

The War of 1812 was followed by a 30-year period of relative peace, with the exception of some border conflicts. The near disaster of the War of 1812 resulted in some improvement to America's homeland defense capability.

States still tightly controlled their militias, and the War Department was not able to centralize or standardize maintenance of the militias, and was limited to supplying

---

[72] Weigley, 55.
[73] Office of the Chief of Military History, "Chapter 6, The War of 1812," 146-147.

training manuals and recommending improvements.[74] However, Congress had learned the importance of a standing Army, and in 1815, voted to establish a peacetime Army strength of 10,000, plus a Corps of Engineers—three times the strength of the Regular Army under the Jefferson administration. The new Secretary of War, William Crawford, convinced Congress to maintain an Army General Staff and add a Quartermaster General to this staff in hopes of more efficiently organizing and supplying future war efforts.[75] Congress also increased the number of Navy warships.[76]

The Coastal Defense Plan

The war also highlighted the need for a coherent national military policy, with the resources to back up that policy. Prior to 1815, defensive fortifications had been built haphazardly without a comprehensive concept for their use or to locate them to support each other. As a result, they were not as effective as they might have been in defending against the British from 1812-1814. After the war, President Madison appointed a Board of Engineers to develop an integrated coastal defense plan, using forts and the Navy as key elements of the design. Their report became America's fundamental maritime defense strategy. It stated that the Navy must be the primary means of defense against invasion, supported by a system of coastal fortifications to protect naval bases.[77]

But as the memory of the war faded, so, too, did Congressional funding for defense. By 1843, 69 fortifications were complete or in progress,[78] but Congress did not appropriate enough money for the entire system of fortifications recommended by the

---

[74] Office of the Chief of Military History, "Chapter 7, The Thirty Years' Peace," in *American Military History* (Washington, D.C.: Office of the Chief of Military History, 1988), 149; on-line, Internet, 30 November 2000, available from http://www.army.mil/cmh-pg/books/amh/amh-07.htm
[75] Office of the Chief of Military History, "Chapter 7, The Thirty Years' Peace," 150.
[76] Weigley, 59.
[77] Weigley, 59-60.
[78] Office of the Chief of Military History, "Chapter 7, The Thirty Years' Peace," 155.

Board of Engineers. Furthermore, the government's immediate priority was now protection of commerce, not building a fleet of warships for major naval engagements. Therefore, rather than the large warships Benjamin Stoddert had envisioned, Congress instead funded small, fast ships to chase pirates.[79] Successive governments through the 1850's maintained this stance (despite some forays into building steam-ships), since large warships served no immediate, compelling national interest.[80]

The Army of the Interwar Years

The Army fared no better. In 1820, Congress cut Army authorizations from 10,000 to 6,000. An 1817 treaty with Great Britain that limited naval armaments on the Great Lakes indicated a Congressional sense that the threat to national security from the north was greatly reduced; by 1846 forts on the Canadian border were neglected and fell into disrepair.[81]

As with the Navy, a large standing Army served no immediate national interest. Threats to our borders came not from European powers with large armies; rather, border conflicts with Native American tribes were the overriding concern.[82] However, our Army was not prepared to fight an enemy that used guerrilla-style tactics. European-style tactics and an Army with heavy logistics requirements were not effective against an enemy that could literally vanish into forests and swamps.[83] In the Second Seminole War (1836-1842), a total of 40,000 Regulars and volunteers fought 7 years to drive the

---

[79] Weigley, 60.
[80] Weigley, 64-65.
[81] Weigley, 60-61.
[82] Weigley, 67.
[83] Weigley, 67-69.

Seminoles out of Florida. The war finally ended by a campaign of extermination, destroying the Seminole's villages and crops.[84]

Professional Officer Development

One success story during this period, however, was the effort to develop a cadre of professional officers. Congress appropriated funds for new buildings, books, maps, and staff for the U.S. Military Academy at West Point in an attempt to improve military education among the officer corps.[85] Sylvanus Thayer, the Superintendent of West Point from 1817 to 1833, worked to develop West Point into the kind of institution that George Washington had envisioned—a military academy dedicated to development of professional officers.[86] However, the nineteenth century obsession with Napoleonic warfare resulted in much emphasis on studying Napoleonic style offensive battles of annihilation. Very little attention was given to the unique American experience of warfare.[87] The West Point curriculum also emphasized engineering and fortifications. The generalship of both the Union and Confederacy in the Civil War were a direct reflection of the principles learned at West Point.[88]

Lessons Learned from the Interwar Years

With the near-defeat of the War of 1812 still fresh in the American collective memory, Congress authorized funding for improvement of U.S. strategic defense capability. But once the memory of the crisis faced in public and government memory, funding for defense quickly eroded, and America was unprepared for the next national defense crisis. This particular pattern of crisis-induced spending, followed by severe

---

[84] Office of the Chief of Military History, "Chapter 7, The Thirty Years' Peace," 159-161.
[85] Office of the Chief of Military History, "Chapter 7, The Thirty Years' Peace," 150-151.
[86] Paret, 412-413.
[87] Paret, 414.

defense cuts, became a hallmark of American homeland defense in years to come, and ensured that America was rarely ready to fight when faced with a national defense crisis.

However, the experience of the War of 1812 did result in Congressional funding for development of a cadre of professional, highly educated military officers. During the interwar period, West Point became an institution dedicated to this goal. Although the curriculum emphasized Napoleonic style warfare, a style of warfare, which was not necessarily appropriate for American capabilities and resources, it began a much-needed tradition of educating and maintaining a core of military professionals who would successfully lead American military efforts in future generations.

The Mexican War

Background

During the years leading up to the Civil War, homeland defense centered on protecting new territories taken from Mexico and Native Americans during this expansionist era. The slogan "Manifest Destiny" epitomized the belief that Americans had the right to expand their territory to the Pacific coast.[89] The government's expansionist designs were naturally met with hostility from the Native Americans and the Mexican government, and the United States government was once again faced with the problem of defending too large a territory with not enough resources. The U.S. Army had been cut to 8500 men after the Second Seminole War[90]—not nearly enough to defend the new territories.

---

[88] Paret, 415-419.
[89] Dupuy and Dupuy, 882.
[90] Office of the Chief of Military History, "Chapter 8, The Mexican War and After," in *American Military History* (Washington, D.C.: Office of the Chief of Military History, 1988), 166; on-line, Internet, 30 November 2000, available from http://www.army.mil/cmh-pg/books/amh/amh-08.htm

The Mexican War

In March 1845, the United States announced annexation of Texas, despite Mexico's threat that such an act would result in war. Once Congress declared war on Mexico in May 1846, Congress funded over 15,000 men to fight. President Polk had hoped for a short, decisive war to establish the U.S. border at the Rio Grande, but, as in previous wars, problems with logistics, mobilization, and inexperienced troops caused the war to drag on for 2 years. However, the superb leadership of Generals Zachary Taylor and Winfield Scott overcame these shortcomings to ensure a decisive victory. Winning the Mexican War not only brought the boundary of U.S. territory to the Rio Grande, but Mexico also ceded vast new territory to the U.S., including present day states of Arizona, New Mexico, Utah, Nevada, portions of Wyoming and Colorado, and upper California.[91]

Strategic Problems in Aftermath of the Mexican War

After the Mexican War, Congress once again reduced the Army's strength to less than 10,000.[92] The reasons were the same as always—fear of a large standing Army and Congressional desire to keep government expenses down.[93]

But the abundance of new territory presented a new problem for the government—how to defend it. Once again, the government's resources failed to match the requirements of its policy. Congress was forced to increase the size of the Army to maintain its claims on the frontier.[94] Defense of the new frontier relied upon a series of outposts on the borders too far apart to effectively coordinate defense. During the 1850s, 90 percent of the Army's 17,000 troops were spread out along a million square mile

---

[91] Office of the Chief of Military History, "Chapter 8, The Mexican War and After," 163-180.
[92] Office of the Chief of Military History, "Chapter 8, The Mexican War and After," 163-180.
[93] Weigley, 71.
[94] Office of the Chief of Military History, "Chapter 8, The Mexican War and After," 180-181.

frontier in the west. Exploration of the new land and protection of an ever-increasing number of settlers who encroached on Native American territory became a full-time job for the Army. During 1857 alone, the Army reported 37 combat engagements with Native Americans.[95] However, defense of the new frontier soon became secondary to the greatest threat yet to the nation—the formation of the Confederacy and the Civil War.

The Mexican War aggravated the tensions between North and South that led to the Civil War. In his memoirs, Ulysses S. Grant pointed out that the Democratic administration wanted to add more pro-slavery states to counterbalance the North's electoral advantage. Texas fit the requirement. But in annexing Texas, the U.S. government only exacerbated the growing cultural and political division between the North and South.[96]

The Civil War

Secession

Before Abraham Lincoln was elected President in 1860, the South and North were already divided culturally, socially, and economically. Tensions over slavery in the new Western states escalated between Southern pro-slavery and Northern anti-slavery states during Buchanan's Presidency. Abraham Lincoln's election in 1860 resulted in South Carolina's secession, quickly followed in February 1861 by Florida, Mississippi, Alabama, Georgia, Louisiana, and Texas. The new Confederacy seized all federal property within its borders, including military forts. When the commander of Fort Sumpter refused to surrender in March 1861, the local Confederate commander started a bombardment. On April 14, 1861, Fort Sumpter's commander surrendered, and Lincoln

---

[95] Dupuy and Dupuy, 950.

called for 75,000 volunteers to fight with the Regular Army to suppress the rebellion.[97] Lincoln's call for volunteers quickly precipitated the secession of Virginia, Arkansas, Tennessee, and North Carolina.[98]

Union Strategic Advantages and Disadvantages

Lincoln hoped for and expected a quick victory, as the Union had many strategic advantages over the Confederacy in military and economic power. The Union had a 5:2 advantage in population, with 22 million people in the Northern states (versus 9 million in the Southern states). The North was both an industrial and agricultural economy with the resources to support the war effort. Its ironworks and munitions plants could internally supply the war effort. The U.S. Navy had only 42 ships in commission when the war broke out, but with its industrial resources quickly built up a Navy that would blockade Confederate harbors.[99]

But the North suffered some strategic disadvantages, as well. America's borders far outreached its ability to protect them. Although the Union had over 16,000 officers in the U.S. Army in 1861, most were guarding America's huge Western frontier, thereby rendering the most experienced soldiers unable to contribute to the war. The Union was forced, as in previous wars, to make up for its manpower deficiency by calling up volunteers, who varied greatly in combat capability.[100]

---

[96] John Keegan, *The Mask of Command* (New York: Viking Penguin Inc., 1987), 184.
[97] Office of the Chief of Military History, "Chapter 9, The Civil War, 1861," in *American Military History* (Washington, D.C.: Office of the Chief of Military History, 1988), 184-189; on-line, Internet, 30 November 2000, available from http://www.army.mil/cmh-pg/books/amh/amh-09.htm
[98] Dupuy and Dupuy, 951.
[99] Dupuy and Dupuy, 952
[100] Office of the Chief of Military History, "Chapter 9, The Civil War, 1861," 189-191.

Confederacy Strategic Advantages and Disadvantages

The South was an agricultural economy, dependent on imports of manufactured goods and exports of its agricultural products for economic livelihood, and had to pay for arms with the agricultural products it exported.[101] But the Confederacy possessed a strategic advantage over the Union. It could win by wearing out the Union. It did not have to conquer the Union to win. The Union, on the other hand, had to conquer the Confederacy to win.[102] The Confederacy hoped to succeed by convincing Europe to join its side and to hold out militarily, exhausting the Union into an agreement.[103]

At the beginning, the Confederacy had the upper hand, tactically. Robert E. Lee's successes at Manassas (both battles), the Seven Days battles, Fredericksburg, and later, his masterpiece, Chancellorsville, were tactical victories, but did not help the Confederacy strategically.[104] These battles achieved attrition on both sides, but the Union was better able to absorb and replace losses than the Confederacy.[105]

Neither "Stonewall" Jackson nor Robert E. Lee was able to force the North to fight on terms advantageous to the South. To win the war, the Confederacy had to "wear out" the North by luring Union forces to fight in the vast open territory of the South, far away from Union supply and communications lines.[106]

Lee protected what both he and the Union forces incorrectly thought to be the Confederacy's center of gravity—the capital at Richmond. The Confederacy's center of

---

[101] Office of the Chief of Military History, "Chapter 9, The Civil War, 1861," 192.
[102] Office of the Chief of Military History, "Chapter 9, The Civil War, 1861," 193.
[103] J. F. C. Fuller, *A Military History of the Western World, Volume III, From the American Civil War to the End of World War II* (New York: Da Capo Press, 1956), 12.
[104] Fuller, 47-48.
[105] Fuller, 47-48.
[106] Keegan, 197.

gravity, however, was not its capital.[107] Rather, the key to defeating the South lay in destroying its supply and transportation systems, for the Confederacy could not long support a war effort without holding key lines of communication or supply hubs.[108]

Scott's "Anaconda Plan"

Winfield Scott was the first on the Union side to recognize this. He proposed a plan to "squeeze" the Confederacy into submission. At first Lincoln paid no attention, because he wanted a quick end to the war. But after the first combat engagements with the Confederacy, Lincoln listened. Scott was the first commander in the Civil War to comprehend the importance of grand strategic design—in this case, the relationship between economic pressure and attack as an overall strategy for winning.[109] He proposed to destroy the Confederacy through a multi-faceted strategy: (1) enforcement of a naval blockade of all major ports, thus preventing agricultural goods from going out, and supplies from coming in, (2) advancing the Army down the Mississippi to split the Confederacy,[110] and (3) contain the Confederate Army in Virginia by advancing on Richmond.[111]

Scott's plan was brilliant. Without the active assistance of European allies, the South had no hope of breaking the North's ever-tightening blockade. The naval blockade of the Confederacy was one of two key strategic moves that helped bring about the end of the Confederacy.[112]

---

[107] Fuller, 48-49.
[108] Office of the Chief of Military History, "Chapter 9, The Civil War, 1861," 193.
[109] Fuller, 12.
[110] Office of the Chief of Military History, "Chapter 9, The Civil War, 1861," 193.
[111] Fuller, 12.
[112] Office of the Chief of Military History, "Chapter 9, The Civil War, 1861," 202-203.

The second strategic stroke was Ulysses S. Grant's capture of Forts Henry and Donelson in Tennessee in February 1862. Like Scott, Grant knew the key to defeating the South was to cut it in half logistically.[113] His capture of these forts opened up the Confederacy to attack from the west.[114]

To make matters worse for the Confederacy, by 1863, the Confederacy realized that neither England nor France would come to its aid—neither was willing to back the losing side. The only hope the Confederates had for winning the war was through a stunning military victory that would demoralize the Union.[115]

Lee's Strategic Mistake: Gettysburg

With his brilliant victory at Chancellorsville in May 1863, Lee had proved the Army of Northern Virginia could defeat Union forces, despite being outnumbered.[116] But Lee also knew that with his limited resources, he did not have the strength to counter the Union in both the West and East. He convinced Jefferson Davis and the Confederate Cabinet that the best strategy would be to attack the Union in the East, forcing the Union Army in Virginia to follow him North (away from the Confederate capital). A great victory in the North could also force the Union to move troops away from the Confederate coastline, and possibly from the Western theater. Finally, he hoped that a victory in the North would enhance the peace movement in the North and cause Lincoln to negotiate.[117]

---

[113] Fuller, 48.
[114] Office of the Chief of Military History, "Chapter 10, The Civil War, 1862," in *American Military History* (Washington, D.C.: Office of the Chief of Military History, 1988), 211-213; on-line, Internet, 30 November 2000, available from http://www.army.mil/cmh-pg/books/amh/amh-10.htm
[115] Timothy H. Donovan, Jr. et al, "The Civil War," in *The West Point Military History Series*, ed. Thomas E. Griess (Wayne, New Jersey: Avery Publishing Group Inc., 1987) 145.
[116] Donovan et al, 142.
[117] Donovan et al, 145-147.

But the loss of his great commander, "Stonewall" Jackson, at Chancellorsville, and the Confederate Cabinet's decision not to give Lee all the reinforcements he requested for his invasion of the North,[118] made the outcome of the most strategically important battle in the East a foregone conclusion. The Confederacy simply did not have the manpower to mount an invasion in the North while trying to protect central Tennessee and Vicksburg in the West. Furthermore, Lee was forced to reorganize his army and appoint new division and corps commanders on the eve of the Battle of Gettysburg.[119] Lee's gamble in switching from a defensive to offensive strategy failed.[120] Lee's loss at Gettysburg cost the Confederacy 28,000 men and its one chance for a strategically significant victory. From that point on, the Confederacy was doomed.

Grant's Strategic Victory: Vicksburg

At the same time, Ulysses S. Grant's forces won a brilliant victory against the Confederate stronghold at Vicksburg, Mississippi—the center of gravity for the Confederacy's logistics. The Union had taken control of the Mississippi River and split the Confederacy.[121]

The Union victory at Chattanooga, Tennessee in July 1863, also gave the Union control of the key railroad hub of the Confederacy. From Chattanooga led the critical railway line to Atlanta, the location of the Confederacy's quartermaster, commissary, and ordnance depots. Capture of Chattanooga gave the Union the means to enter the interior of Georgia and the rest of the Deep South.[122]

---

[118] Donovan et al, 146-147.
[119] Donovan et al, 147, 164-165.
[120] Donovan et al, 165.
[121] Office of the Chief of Military History, "Chapter 11, The Civil War, 1863," in *American Military History* (Washington, D.C.: Office of the Chief of Military History, 1988), 237-241; on-line, Internet, 30 November 2000, available from http://www.army.mil/cmh-pg/books/amh/amh-11.htm
[122] Donovan et al, 174.

Grant's Final Plan

As welcome as these victories were for the Union, they were not coordinated nor designed to complement each other. Grant realized that a comprehensive strategic plan was needed to finish off the Confederacy.[123] Grant's strategy was simple—finish off the Confederate Army and destroy the Confederacy's ability to wage war.

After Lincoln appointed him General-in-Chief of the Union Army, Grant insisted that the Union armies must act in concert under his orders to bring the full force of Union capabilities against the Confederate Army. He also began a war of attrition to destroy the Confederacy's war-making capacity and demonstrate that the Confederate Army could not protect Southern territory.[124] General Sherman's "march to the sea" through Georgia and thence through the Carolinas did just that, laying waste to all farms, railways, and storehouses in his army's path. Sherman's actions were a prelude to the "total war" concept of the twentieth century, when strategic bombardment and economic warfare became integral parts of wartime strategy.[125]

Civil War Lessons Learned

The Civil War started as previous American wars had, with little thought to a viable grand strategy and a military unprepared for the task at hand. The Union did not use its strategic advantages (more men and material resources) to great effect at first, until Winfield Scott produced a plan that would use the North's strengths against the South's weaknesses (an agricultural economy, dependent on long, vulnerable lines of communication and logistics, and ports to obtain basic material needs). And despite the

---

[123] Office of the Chief of Military History, "Chapter 12, The Civil War, 1864-1865," in *American Military History* (Washington, D.C.: Office of the Chief of Military History, 1988), 262-263; on-line, Internet, 30 November 2000, available from http://www.army.mil/cmh-pg/books/amh/amh-12.htm
[124] Donovan et al, 197, 224.

critical Union victories at Chattanooga and Vicksburg, only after these uncoordinated victories did Grant obtain the authority for all Union forces to act under his orders, to ensure a coordinated grand strategic design for final defeat of the Confederacy.

The South wasted its lesser resources in a series of strategically meaningless battles, and erroneously tried to protect its Capital at Richmond, failing to realize its real vulnerability lay in access to resources through its ports and railways. Furthermore, instead of sticking to a strategy of wearing out the Union (a necessity for a weaker force), Lee gambled and lost in an attempt to create a strategically important victory at Gettysburg.

Both sides initially repeated the same mistakes that American forces had committed in previous wars. It took the strategic genius of Scott and Grant to pull together a plan to win the war. Unfortunately, the American habit of developing a workable plan after the war already started continued until well into the next century.

Post-Civil War

Post-Civil War Defense Policy

The American policy of severely cutting the armed forces after war continued after the Civil War. Furthermore, the government continued its habit of spreading the military too thinly to cover the required missions. The Army was cut back severely after the Civil War, but it still had the two enormous missions of policing the Western frontier and the Southern states during Reconstruction.[126] America's incessant expansion to the West brought about numerous fights with Native Americans. The Army fought over 900

---

[125] Office of the Chief of Military History, "Chapter 12, The Civil War, 1864-1865," 273.
[126] Weigley, 167.

actions with Native Americans from 1864 to 1898.[127] Furthermore, the French sent troops into Mexico in 1864 and established a puppet government, necessitating a 52,000-man U.S. Army show of force in Texas in late May 1865.[128]

Based on these strategic requirements, General Grant argued the necessity of maintaining an 80,000-man army, but to no avail. By 1876, Congress cut Army authorizations to 27,442—and funded only that number until the outbreak of the Spanish-American War.[129]

And the Navy, after pioneering the use of ironclads (thus rendering wooden warships obsolete during the Civil War)[130] now returned to using squadrons of steam-powered wooden ships in its patrols.[131] Instead of maintaining a viable fleet of modern warships, the government returned to its pre-War of 1812 passive-defense policy of relying on coastal fortifications for defending coastal cities against naval attack.[132]

Origins of the National Guard

One aspect of homeland defense that received more attention was the militia. Volunteer militias were used frequently during the 1870's to quell labor unrest. State governments recognized the value of state militias, and between 1881 and 1892, every state began provisioning their militias. Most states also gave their militias a new name:

---

[127] Dupuy and Dupuy, 991.
[128] Office of the Chief of Military History, "Chapter 13, Darkness and Light: The Interwar Years, 1865-1898," in *American Military History* (Washington, D.C.: Office of the Chief of Military History, 1988), 281-282; on-line, Internet, 30 November 2000, available from http://www.army.mil/cmh-pg/books/amh/amh-13.htm
[129] Office of the Chief of Military History, "Chapter 13, Darkness and Light: The Interwar Years, 1865-1898," 282.
[130] Dupuy and Dupuy, 902-903.
[131] Weigley, 167-168.
[132] Weigley, 167.

the National Guard. By 1898, it was the reserve force for the Regular Army.[133] Despite attention from the states, however, Congress did little to improve the National Guard's capability, so the new National Guard units did not receive the needed equipment, training, or funding they needed to become truly viable military reserve units.[134]

Post-Civil War Threat Environment

These policies directly reflected the threat environment of the era. No country was able to threaten America with an 1812-style blockade since steam power made ships dependent on coal supplies, thereby limited their ability to project power across the ocean. European states were more concerned with protecting their colonial interests and resolving conflicts between themselves. America could now concentrate on its own domestic security, and its defense policies reflected this fact.[135]

Analysis of American Homeland Defense: The Colonial Era Through The Post-Civil War Era

In reviewing the period from the Colonial Era through the Post-Civil War Era, a clear, repetitive pattern of American homeland defense strategy and policy emerges:

    a.   Suspicion of large standing armies and desire for economy in defense expenditures resulted in massive cuts to regular forces after every major conflict. As a result, Americans were forced to rely on thousands of untrained, undisciplined non-professional volunteers and state militia forces to fill the ranks each time war broke out. The cost of this policy was especially apparent at the outbreak of the War of 1812, when

---

[133] Office of the Chief of Military History, "Chapter 13, Darkness and Light: The Interwar Years, 1865-1898," 287.

America had neither the ground nor naval forces to successfully defeat the British without foreign help.

b. The colonial and state militias were in dire need of centrally standardized and mandated standards for training, equipment, and funding. However, each colony and state jealously guarded its prerogatives over its own militia forces. As a result, the efficiency and combat-worthiness of the militias varied greatly between the colonies and states. When wars erupted, the militias were not sufficiently practiced in drill, discipline, or tactics to compete with regular enemy forces.

c. The American government consistently tried to defend too much territory with not enough forces. Outside help (from the French) was needed to prevent defeat in both the Revolutionary War and War of 1812, and the U.S. Army was never able to effectively defend the new frontier after the Mexican War. During the Civil War, the Confederate government never fully realized this strategic predicament, spreading its limited forces too thinly to achieve strategic effect.

d. In each of the major wars, Americans wasted valuable men and resources early in the wars on tactical battles with little or no strategic value due to failure to develop a comprehensive, overarching strategy that could be supported with available resources. In the Revolutionary War, it took Washington two major defeats to realize that his forces

---

[134] Office of the Chief of Military History, "Chapter 13, Darkness and Light: The Interwar Years, 1865-1898," 287.

could never win against the British using a conventional European offensive strategy. His armies were forced to use a more unconventional approach—wearing out British forces with hit and run tactics. The War of 1812 was a particularly egregious example, for one objective of the war was to force the British to respect American naval rights on the high seas. But we did not have the military capability to achieve this objective, and the British successfully blockaded our harbors. The land strategy was no better; the failed invasion of Canada was yet another example of a political objective which far outreached our capability. The Civil War started off no better, with both sides wasting their resources on useless tactical battles instead of developing a successful grand strategic design to attack the enemy's true center of gravity. The strategic geniuses of the Civil War—Scott, Grant, and Sherman—were the first to realize the concept of a grand, overarching strategy using economic, political, and military instruments of war in a strategic plan to squeeze the Confederacy into defeat. The Confederacy, on the other hand, repeated the strategic blunders of the Revolutionary War and War of 1812 by trying to defend too many places at the same time with too few resources. Furthermore, Lee's greatest strategic blunder, switching from a defensive strategy to an offensive strategy without the resources assure victory, heralded back to the same mistakes Washington made in the Revolutionary War.

---

[135] Weigley, 168.

e. Both the Revolutionary War and War of 1812 highlighted the critical need for professional military officers schooled in the science and art of war. Without a will-educated, experienced cadre of officers, the military could not hope to develop viable military strategy to support the government's homeland defense requirements. Establishment of the U.S. Military Academy at West Point ensured more attention to development of professional officers, but the curriculum stressed study of warfare doctrine that was not necessarily applicable to American defense needs or capability. Furthermore, the entire period highlighted the need for militarily literate civilian leaders who could comprehend homeland defense requirements and would fund a standing force capable of executing America's policies and objectives.

Chapter 3

American Homeland Defense: The Expansionist Era Through World War II

The country must have a large and efficient army, one capable of meeting the enemy abroad, or they must expect to meet him at home.

—The Duke of Wellington

The Expansionist Era: 1865-1898

Background

Through the end of the Civil War, America was self-absorbed in its defense policies. For 250 years, the country had concentrated first on its own survival, then on expanding its hold over the rest of the continent between Mexico and Canada.[136] But by the closing years of the 19th century, America possessed the continent from the Atlantic to the Pacific, and the nation started to become involved in matters outside its own borders.

America's rapidly expanding industrial and economic base needed more markets to increase capital. Observing the economic and political benefits the European powers gained from colonial expansion, America joined in the game.[137] The purchase of Alaska in 1867, development of a coaling station in Samoa for our fleet of steam-powered ships,[138] and the deposition of Hawaii's Queen by a group of American businessmen and

---

[136] Office of the Chief of Military History, "Chapter 15, Emergence to World Power 1898-1902," in *American Military History* (Washington, D.C.: Office of the Chief of Military History, 1988), 319; on-line, Internet, 30 November 2000, available from http://www.army.mil/cmh-pg/books/amh/amh-15.htm
[137] Weigley, 170.
[138] Office of the Chief of Military History, "Chapter 15, Emergence to World Power 1898-1902," 319.

imposition of a new government in 1893, were manifestations of America's new expansionism.[139]

However, America's expansionist policies required the military power to back them up. But, as in the aftermath of previous wars, military forces were severely cut after the end of the Civil War. Between 1864 and 1870, the U.S. Navy was cut from a total of 700 ships carrying over 5000 total guns to 200 ships carrying only 1300 guns.[140]

Alfred Thayer Mahan and Control of the Sea

The great American naval strategist and theorist, Alfred Thayer Mahan, insisted that control of the sea was essential to economic and territorial expansion. According to Mahan, control of the sea could only be achieved by building a superior naval force to destroy the enemy fleet. He flatly rejected historic American naval policies of building ships for commerce raiding, since the cruisers used for commerce raiding could never achieve control of the sea.[141]

Mahan was also one of the first the realize America's historic strategic dilemma of too much territory to protect and not enough resources to protect it. America had to both protect and project power from two separate coasts. The Navy had to control interior lines of communication throughout the Caribbean, develop bases in the Pacific, and build a U.S.-controlled canal across the isthmus of Panama to enable our Navy to move quickly between oceans to protect our lines of communication.[142] He also rejected the traditional American government policy of dispersing its forces to cover too much territory; rather,

---

[139] Howard Cincotta, ed., *An Outline of American History* (Washington, D.C.: United States Information Agency, May 1994), n.p.; on-line, Internet, 22 December 2000, available from http://usinfo.state.gov/usa/infousa/facts/history/ch7.htm#ambivalent.
[140] Paret, 469.
[141] Weigley, 173-176.
[142] Weigley, 177-178.

he advocated concentrating the battle fleet for its primary mission of defeating an enemy fleet.[143]

At first, Congress did not fully appreciate the soundness of Mahan's ideas. In 1890, Congress only authorized 3 battleships with such limited fuel endurance (5000 nautical miles) that made offensive capability impossible.[144] But by 1893, with the coup in Hawaii quickly followed by the new government's request for annexation, and the concerted efforts of pro-Mahan Secretaries of the Navy Hilary Herbert and Benjamin Tracy,[145] Congress authorized construction of six more battleships by 1896.[146]

The U.S. Navy in the Expansionist Era

The Navy also enhanced homeland defense capability during this era by developing a cadre of highly educated professional officers. The Navy established the Naval War College at Newport, Rhode Island in 1885. This new institution trained officers in war theory and strategy, with special emphasis on Mahan's theories.[147]

The U.S. Army in the Expansionist Era

Thus, the combination of Mahan's genius, creation of a cadre of well-educated Naval officers, and selection of militarily literate, supportive Navy civilian leadership, created a Navy ready for war with an enemy close to home.

The Army, however, did not fare as well as the Navy during this era. Congress had cut Army authorizations to slightly over 26,000 men after the Civil War and had dispersed its limited manpower across the country in companies and battalions. With such small units scattered so widely, the Army never had the opportunity to train in units

---

[143] Weigley, 179.
[144] Weigley, 182-183.
[145] Office of the Chief of Military History, "Chapter 15, Emergence to World Power 1898-1902," 322.
[146] Weigley, 183.

larger than regiment size. Furthermore, the Army had no mobilization plan and never trained in joint operations with the Navy.[148] The militias were in even worse shape. Training for state militias was practically non-existent. States did not provide adequate funding or equipment for militia units.[149] Thus, unlike the Navy, the Army was unprepared for the next test of American defense capability—the Spanish-American War.

The Spanish-American War

Background

The Spanish-American War can be traced to a number of causes. American nationalistic and expansionist fervor certainly contributed to this new foray into fighting well beyond America's borders.[150] America wanted absolute control of its economic and trade interests in the Caribbean and the Pacific—and Spain was in the way.

At the time, Cuba was a Spanish colony. When Cubans rebelled against Spain's repressive colonial policies in 1868 and 1895, many Americans sympathized with the Cuban rebels. When Spain instituted a policy of forcing elderly people, women, and children into detention camps, where many starved and died of disease, American press reports of these policies whipped up fury against Spain. Furthermore, American expansionists and business opportunists had not failed to notice the economic potential of Cuba and its strategic location.[151]

---

[147] Office of the Chief of Military History, "Chapter 15, Emergence to World Power 1898-1902," 322.
[148] Office of the Chief of Military History, "Chapter 15, Emergence to World Power 1898-1902," 322-323.
[149] Brian Blodgett, "The Difficulties in the Formation of V Corps for the Spanish-American War," 1-3; on-line, Internet, 27 October 2000, available from http://members.tripod.com/Brian_Blodgett/V_Corps_1898.html.
[150] Office of the Chief of Military History, "Chapter 15, Emergence to World Power 1898-1902," 319-320.
[151] Office of the Chief of Military History, "Chapter 15, Emergence to World Power 1898-1902," 319-321.

The mysterious explosion of the *Maine* in February 1898 in Havana's harbor was the catalyst for war. An investigating commission determined that an external explosion had destroyed the Maine. With the sensationalistic press blaming the explosion on Spain and inciting public furor, President McKinley ordered a blockade of Cuba, and in April 1898, Congress declared war on Spain.[152]

Strategic Weaknesses in the Spanish-American War

When Congress declared war on Spain, many volunteer militia units were not deployable due to lack of equipment, poor training, and lack of organization.[153] Worse, the War Department had not prepared for the possibility of equipping over 200,000 Regulars and volunteers on short notice. There was no coordinated plan for mobilization, nor did funds exist to equip the Army.[154] The supply system was so poor that deficiencies in basic supplies were not corrected.

Furthermore, no real strategy existed for employing American forces. There had been no cooperation between Congress and the War department on coordinating military force structure and training with foreign policy objectives.[155]

America went to war with the idea of forcing Spain to leave Cuba by using a naval blockade while Cuban guerrillas harassed Spanish occupying forces on land. No one planned on any large land engagements with the Spanish on Cuba; American Army forces would simply occupy Cuba after the Spanish departed.[156]

However, the public, fueled by inflammatory press stories, wanted immediate action against the Spanish. So, despite the fact that the Army was poorly equipped, poorly

---

[152] Office of the Chief of Military History, "Chapter 15, Emergence to World Power 1898-1902," 322.
[153] Blodgett, 1-3.
[154] Office of the Chief of Military History, "Chapter 15, Emergence to World Power 1898-1902," 323-324.
[155] Office of the Chief of Military History, "Chapter 15, Emergence to World Power 1898-1902," 323-326.

trained and disorganized, had no real plan of operations, nor any intelligence on the strength and disposition of Spanish forces, the Secretary of War ordered the Army to prepare for an invasion.[157]

Naval Strategy in the Spanish-American War

When the war began, fortunately it was the Navy that bore the brunt of the important operations. The U.S. Pacific Fleet was able to concentrate its forces and achieved a great strategic success in the Philippines. The Spanish fleet at Manila had to be destroyed to prevent it from aiding Spain in the Caribbean and threatening American harbors and shipping.[158] Achieving complete surprise, Commodore George Dewey's Asiatic Squadron swiftly and decisively defeated the entire Spanish Fleet in a Mahan-style all-out battle on April 30, 1898, at Manila harbor within hours of engaging the enemy.[159] Dewey then blockaded Manila until American troops arrived 2 months later.[160]

But interference from Washington disrupted operations in the Caribbean. Admiral Simpson's plan for a full-strength Atlantic naval squadron to blockade Cuba and intercept the Spanish fleet was thwarted when rumors reached the American public that the Spanish fleet was approaching the Atlantic Coast. The Department of the Navy caved in to public demands and withheld some of the war ships to protect the Atlantic Coast, thereby dispersing Admiral Simpson's fleet and leaving it unable to enforce a complete blockade of Cuba. As a result, the Spanish Fleet in the Atlantic was able to slip past the blockade and into Santiago harbor on May 19, 1898.[161]

---

[156] Office of the Chief of Military History, "Chapter 15, Emergence to World Power 1898-1902," 323-325.
[157] Office of the Chief of Military History, "Chapter 15, Emergence to World Power 1898-1902," 323-326.
[158] Cincotta, 3-4.
[159] Office of the Chief of Military History, "Chapter 15, Emergence to World Power 1898-1902," 326-327.
[160] Dupuy and Dupuy, 994.
[161] Office of the Chief of Military History, "Chapter 15, Emergence to World Power 1898-1902," 325-327.

Land Operations in the Spanish-American War

Simpson requested land forces to assist his fleet in taking Santiago. The War Department selected V Corps, the only Army unit anywhere near ready for action, to conduct the land campaign in Cuba. But lack of skilled direction in loading needed supplies, only one pier for loading ships at the marshalling area, not enough ships to carry the men and equipment, and no regard of combat zone unloading priorities delayed departure of V Corps for 2 weeks. When V Corps did land in Cuba, confusion reigned. Captains of the troop ships refused to get close to shore to unload horses, men, and supplies, slowing the landing of the force. The 36,000 Spanish troops in Santiago Province could have prevented the force from landing, but for some reason did not. So some 17,000 American troops and about 5000 additional Cuban insurgents landed unopposed at Santiago.[162]

Failure to Coordinate Land and Naval Operations

Initial land operations against the Spanish in Cuba were characterized by poor command, control, coordination, and reconnaissance. The V Corps Commander also could not convince the Navy to launch a coordinated assault on Santiago Bay to complement land operations. Heat, tropical disease, difficult terrain and lack of experienced leadership slowed the now-famous assaults at San Juan and El Caney on July 1, and although they were eventually successful, 1700 men were lost in the process.[163] Two days later, the American fleet destroyed the Spanish fleet as it tried to escape Santiago Bay, and on 17 July, the Spanish Commander at Santiago surrendered.[164]

---

[162] Office of the Chief of Military History, "Chapter 15, Emergence to World Power 1898-1902," 328-332.
[163] Office of the Chief of Military History, "Chapter 15, Emergence to World Power 1898-1902," 322-334.
[164] Dupuy and Dupuy, 995.

Although the war ended favorably for America, incompetent Spanish military leadership contributed to our victory. The outcome might have been very different if the enemy had chosen to defend Cuba more aggressively.

Spanish-American War Lessons Learned

For the Army, the war highlighted the same weaknesses in American homeland defense capability that historically developed in the years between our major conflicts. Dispersal of too few forces in too small units across too much territory made coordinated training between units impossible. Failure to train, organize, and properly equip a large enough core of experienced Regular Army and militia, and failure of the Army and Navy to train for joint operations all contributed to the Army's fiasco in planning, organizing, and executing its mission in Cuba.

Conversely, the Navy had the advantage of Mahan's strategic vision and supportive Navy Secretaries. It proved the long-term necessity of maintaining a well-trained, well-equipped force large enough to achieve decisive victory, even during years of relative peace.

The Interwar Years Through World War I: 1898-1918

Strategic Requirements Versus Resources

The Navy's incredible success during the Spanish-American War brought about unanticipated strategic consequences; namely, the requirement to protect new American possessions abroad, including the Philippines, Hawaii, Puerto Rico, and Guam.[165] As a result, about one-third of the Regular Army was stationed overseas to protect American possessions. Troops served in the Philippines to suppress insurrection; Regulars were

---

[165] Office of the Chief of Military History, "Chapter 15, Emergence to World Power 1898-1902," 336.

48

also stationed in Alaska, Hawaii, and China. However, as usual, the military did not have enough resources to execute this forward-basing policy and ensure adequate U.S. territorial defense. Between 1902 and 1911, the Army had only 75,000 men to cover all of its requirements.[166]

The Dick Act: Federalizing the Militia

In an attempt to remedy the military manpower shortage, Congress revised the obsolete Militia Act of 1792, and passed the Dick Act in 1903.[167] The Dick Act federalized the militia, recognizing the National Guard as the organized militia, and provided it with federally funded equipment and training, and prescribed regular training periods.[168]

The Beginning of U.S. Strategic Defense Planning

1903 also saw the beginnings of U.S. strategic defense planning. The Joint Army and Navy Board, created in 1903, was the first U.S. military inter-service planning organization. It developed policy for U.S. territorial defense, as well as defense of the Western Hemisphere (in accordance with the Monroe Doctrine). The Joint Board recommended establishment of bases in strategic locations, advantageous placement of forces, and preparation of war plans for defense of U.S. possessions and territory.[169] The Joint Board's two most important war plans for protecting American territory and possessions, first developed during the early years of the Wilson administration, were ORANGE and BLACK. Both were conceptually flawed and unrealistic.[170]

---

[166] Office of the Chief of Military History, "Chapter 15, Emergence to World Power 1898-1902," 350.
[167] Office of the Chief of Military History, "Chapter 15, Emergence to World Power 1898-1902," 351.
[33] Dougherty, 969.
[169] Weigley, 100-101.
[170] Michael J. McCarthy, "Lafayette, We Are Here: The War College Division and American Military Planning for the AEF in World War I" (master's thesis, Marshall University, 1992), 3-4; on-line, Internet, 25 October 2000, available from http://mccarthy.marshall.edu/thesis/aef_2.txt.

War Plans ORANGE and BLACK

War Plan ORANGE, completed in 1914 (and continually revised throughout the years between World War I and World War II), promulgated a scenario in which the United States would defend Manila against a Japanese invasion by a naval battle within 1200 miles of the Philippines.[171] The absurdity of this plan was the fact that Japan could land 50,000-60,000 men in the Philippines within about a week, and about 300,000 men in a month.[172] However, it would take the U.S. fleet over 60 days to cross the Pacific to reach the 17,000-man U.S. garrison in the Philippines.[173] Furthermore, U.S. Navy capabilities in 1914 (and, indeed, after World War I, as well) could not support such a plan. The U.S. Navy did not have enough auxiliary ships (colliers and oilers) to support a fleet Pacific crossing in such a time frame.[174] However, this plan remained in force, even through the Post-World War I era, in part because it justified the Navy's budget and claim to resources[175].

War Plan BLACK was a naval strategy to defend against a German fleet attack in the Caribbean or on the American homeland. The plan called for the US fleet, stationed in Cuba and Puerto Rico, to defend against a German naval attack 500 miles from shore, thereby preventing landing of German troops on U.S. homeland or territories. As with ORANGE, the logistical problems of this strategy made the plan's viability questionable.[176]

---

[171] McCarthy, 3.
[172] Allan R. Millett and Williamson Murray, ed., *Military Effectiveness Volume II: The InterWar Period* (Boston: Unwin Hyman, Inc., 1988), 78-79.
[173] McCarthy, 3.
[174] McCarthy, 3.
[175] Millett and Murray, 82.
[176] McCarthy, 3.

The Wilson Administration: Homeland Defense Policy

Isolationist and pacifistic sentiments among both the U.S. public and the Wilson administration ensured little attention to addressing the U.S. military's severe supply and logistical problems for each of the military services. For example, when World War I broke out in 1914, the Chief of Staff of the Eastern Department had no maps of the European theater and no money to buy any. He had to request a colleague to search the War College for copies of needed maps.[177]

But from 1914-1917, the sinking of the *Lusitania*, an effective Allied propaganda campaign against Germany, Germany's violation of Belgian neutrality, German espionage in the U.S., and Germany's continued submarine attacks on Allied shipping increased public pressure on the Wilson administration to take new measures to enhance military readiness and national defense capability, despite his efforts to keep the U.S. out of the war.[178]

Mexican bandit Francisco "Pancho" Villa's border raid in Columbus, New Mexico in March 1916 gave further impetus to force Congress and the President to improve homeland defense capability. Although American forces garrisoned at Columbus defeated the attacking force,[179] this raid highlighted the need for U.S. forces to guard the

---

[177] McCarthy, 4.
[178] Office of the Chief of Military History, United States Army, "Chapter 17, World War I: The First Three Years," in *American Military History* (Washington, D.C.: Office of the Chief of Military History, 1988), 364-366; on-line, Internet, 26 October 2000, available from http://www.army.mil/cmh-pg/books/amh/amh-17.htm.
[179] James P. Finley, "Buffalo Soldiers at Huachuca : Villa's Raid on Columbus, New Mexico," *Huachuca Illustrated* 1, (1993): n.p.; on-line, Internet, 9 November 2000, available from http://www.ukans.edu/~kansite/ww_one/comment/huachuca/HI1-12.htm.

Mexican border, as well as prepare for the eventuality of war. As a result, Congress took action to enhance America's defenses.[180]

National Defense Act of 1916

The result was the National Defense Act of 1916. This Act increased the size of the military and granted new powers to the President. Congress increased peacetime Regular Army strength to 175,000 men and increased the National Guard to over 400,000, making the Guard the core of the citizen army. Furthermore, this Act made the Guard subject to Presidential call-up, increased funding for the Guard, and mandated federal standards for organization and training.[181]

More importantly, this Act recognized the fact that homeland defense had quantitatively changed since the previous century, and that mobilizing for war required the cooperation of every major sector of the American economy. As a result, this Act created the Council of National Defense. The Council was composed of leaders in industry and labor, as well as the Secretaries of War, Navy, Interior, Agriculture, Commerce, and Labor—in essence, people who had the power to mobilize the economy to support a major war.[182] This Act also empowered the President to order defense materials and force industry compliance with his orders.[183]

America's New Homeland Defense Requirements

No longer was defense simply a matter of adequate military forces and coastal fortifications—defense concerned the entire American economy in realization of the fact

---

[180] Office of the Chief of Military History, United States Army, "Chapter 17, World War I: The First Three Years," 367-368.
[181] Office of the Chief of Military History, United States Army, "Chapter 17, World War I: The First Three Years," 367.
[182] Office of the Chief of Military History, United States Army, "Chapter 17, World War I: The First Three Years," 367-368.
[183] McCarthy, 14.

that the war in Europe directly affected America. American lives were lost in German submarine attacks on Allied ships, and our economy was threatened by Germany's attacks on shipping. Furthermore, Germany's political intrigues against the U.S. affected our economic and political influence. America was now an industrial and colonial power, and its economic and political ties to European nations inexorably drew it into World War I, despite President Wilson's fruitless efforts to remain neutral.

America Enters World War I

Germany's proclamation of unrestricted submarine warfare in January 1917, coupled with secret negotiations with Mexico and Japan for a potential German-Japanese-Mexican alliance, ended any remaining hopes of American neutrality. Publication of the Zimmerman Note, which proposed a Mexico-Germany alliance in event of war between Germany and the U.S., recapture of Mexico's lost territories in the U.S., and the proposal that Mexico request Japan to join in an alliance with Germany, further inflamed American anger against Germany. [184] Finally, President Wilson asked Congress to declare war on Germany April 2, 1917, after Germany sank 4 more American ships, killing 15 Americans in the weeks following exposure of the Zimmerman Note.[185]

Homeland Defense Strategy and Resources in World War I

But as usual, Congressional limits on funding and manpower during a decade of relative peace left the military in no shape for deployment into a full-scale war. The Army's arsenal consisted only of approximately 890,000 Springfield rifles. Only 210,000 men were in the Army—and this included National Guardsmen called up for federal duty on the Mexican border. The Army had not one unit of division size (28,000

---

[184] Dupuy and Dupuy, 1060.

men),[186] so one was hastily put together from several regiments, along with some Reserve officers for staff positions.[187] Congress passed the Selective Service Act in May 1917 to provide the hundreds of thousands of additional troops needed in Europe.[188]

By contrast, the Navy was able to aid the Allies from the start. Rear Admiral William Sims convinced the British to try a convoy system to counter the devastating effects of German U-boats. In May, destroyers and other armed U.S. ships began escorting merchant ships across the Atlantic, significantly reducing shipping losses from German U-boats.[189]

Problems in Mobilizing Industry for Homeland Defense in World War I

The most important break from American homeland defense tradition was the mobilization of American industry and populace in an all-out effort to support the war. The results, especially in the first months after America's entry into the war, were inauspicious. The shipbuilding industry speedily began building ships to counteract losses from U-boats, but ports were so clogged with supplies, that British shipping had to take much of the cargo. The Council of National Defense established a War Industries Board to coordinate Army and Navy purchases and convert industrial plants to military use. However, this late attention to the needs of equipping soldiers forced the Army to train with obsolete and sometimes fake weapons. As a result, the U.S. Army had to rely on the Allies for most weapons (except rifles) and many basic supplies such as blankets.

---

[185] Office of the Chief of Military History, United States Army, "Chapter 17, World War I: The First Three Years," 370.
[186] Office of the Chief of Military History, United States Army, "Chapter 17, World War I: The First Three Years," 375.
[187] Office of the Chief of Military History, United States Army, "Chapter 17, World War I: The First Three Years," 372-373.
[188] Office of the Chief of Military History, United States Army, "Chapter 17, World War I: The First Three Years," 374.

Railways became so clogged and slowed with wartime transport requirements that the government took them over and ran them.[190]

America's difficulties in mobilizing for World War I were caused by some of the same attitudes and policy shortcomings that caused difficulties in mobilizing for previous wars. During the two decades of relative peace, Congress did not provide enough funding for military forces to maintain enough trained, equipped troops to quickly mobilize for war. War Plans ORANGE and BLACK, the primary plans to execute American homeland defense policies, were completely unrealistic, given the limitations Congress placed on military funding and resources. When the Wilson administration did finally realize the extent to which the American industry would have to be mobilized to support American involvement in World War I, organization and coordination of production and transportation were inefficient, delaying adequate logistical support for military forces. The same policy shortcomings plagued American defense efforts despite two centuries of experience.

The Interwar Years Through World War I: Lessons Learned

A key lesson learned was the fact that U.S. homeland defense would never again be a strictly domestic issue. Despite President Wilson's efforts to remain neutral, the crisis that enveloped Europe when World War I began directly affected the safety, prosperity, and security of America. German espionage, attacks on merchant shipping, which killed Americans, and the fate of our key economic and political allies, now directly affected

---

[189] Office of the Chief of Military History, United States Army, "Chapter 17, World War I: The First Three Years," 374.
[190] Office of the Chief of Military History, United States Army, "Chapter 17, World War I: The First Three Years," 376-378.

America's vital economic and political interests. Defense of European allies now became a critical aspect of American homeland defense.

But the government continued to make some of the same defense policy blunder as it did in the previous two centuries' wars by requiring the military to defend too much with too few resources in its first war plans. But although the original plans (ORANGE and BLACK) were not logistically supportable, they were important first steps in developing a long-term defense planning capability later in the 20th century.

Congress also took some important legislative steps to ease transition from peace to war. Congress increased the size of the Regular Army, and through the Dick Act and the National Defense Act of 1916, the government federalized the Militia and increased it to over 400,000.

Despite these increases in manpower, the U.S. Army was not ready when America entered World War I. The government had not provided enough basic supplies (such as rifles and blankets) for all of the troops, and Regular Army units were still undermanned at the outbreak of war.

The government's attempt to ease wartime transition by creating the Council of National Defense did not begin auspiciously, either. As in previous American war efforts, the supply system could not handle the sudden increase in material, thanks to last minute planning and an inefficient transport system. Ports became clogged, and critically needed supplies were late in reaching American troops.

Essentially, America encountered the same problems in mobilizing for World War I that it did in previous wars, but the President and Congress had at least recognized and attempted to alleviate the problems of transitioning to war before America actually entered World War I.

Interwar Years: 1919-1941

Retrenchment and Isolationism

Homeland defense capability declined significantly in the aftermath of World War I. The U.S. military found little public or Congressional support for maintaining capability to effectively carry out wartime missions.[191] This period coincided with a resurgent isolationism in the bitter aftermath of World War I. President Hoover reinforced this attitude, stating, "We shall enter into no agreements committing us to any future course of action or which call for the use of force to preserve peace."[192]

One bright spot in the homeland defense arena was in improvement in procurement planning and procedures. The fiasco of producing and distributing goods to execute operations in World War I led to reorganization of the War Department's supply and purchasing procedures through the National Defense Act of 1920. This Act created the Assistant Secretary of War, who was responsible for procurement planning. In 1924, the Army Industrial College began its program of training officers in all aspects of budget and logistics.[193] Unfortunately, improvements in procurement were not matched by coordination of strategic goals and requirements to logistics capability. No attempt was made to match military planning and strategy to actual wartime production capacity.[194]

Erosion of Military Readiness

---

[191] Ronald Spector, "The Military Effectiveness of the US Armed Forces, 1919-39," in *Military Effectiveness Volume II: The Interwar Period*, ed. Allan R. Millett and Williamson Murray (Boston: Unwin Hyman, 1988), 70.
[192] Spector, 70-71.
[193] Spector, 80-81.
[194] Spector, 80-81.

Additionally, the government's policy of fiscal restraint ensured the military's actual capability could not meet mission requirements. Although the National Guard remained about 200,000 strong, its twice-monthly training rituals were wholly inadequate to prepare men for the complexities of modern warfare.[195] From the 1920's through the early 1930's, funded Army authorizations usually remained at less than 135,00—less than one-fourth the strength the War Department wanted.[196] Soldiers who remained in the military were reduced in grade and many were forced to live in uninhabitable quarters. Training in armor units was inadequate because budgets did not provide for enough fuel to maintain armor unit proficiency.[197] Military readiness further eroded through the Army's own organizational policies. The Army insisted on maintaining its 1919 structure of 9 divisions (all incomplete) instead of restructuring to maintain smaller, full-strength units.[198]

The Navy also suffered numerous setbacks during the interwar period. The 1922 and 1930 international disarmament agreements severely limited the number of U.S. warships, tonnage, and armament. These agreements determined and limited Navy fleet composition not by policy or strategy requirements, but by political negotiation. Furthermore, the 1922 agreement disallowed any new construction of bases or fortifications in our Asian possessions, including the Philippines and Guam. In essence, this agreement made it impossible to successfully execute a naval war against Japan.[199] Despite the Navy's intense lobbying effort, Congress refused to allow construction of a

---

[195] Spector, 72.
[196] Spector, 71.
[197] Spector, 72.
[198] Spector, 72.
[199] Spector, 72-73.

naval base on Guam,[200] even after Japan invaded Manchuria in 1931 and denounced the disarmament agreements, giving two years notice of withdrawal in 1934.[201] To make matters worse, the government remained committed to War Plan ORANGE as its primary strategic defense plan, even though the Navy did not have the ships or manpower to execute it. The plan was conceptually flawed in that it gave the Japanese a clear advantage in location and lines of communication, and Congressional refusal to develop Pacific naval bases made it logistically unsupportable.[202]

The Army Air Corps and Homeland Defense

Another major problem concerned planning for coastal defense. Coastal defense was a basic homeland defense mission, and previous to this era, the roles of the Army and Navy in coastal defense were well defined. The Navy was to engage hostile forces at sea, and the Army would engage any forces attempting to land. But with the advent of the Army Air Corps, roles and missions in coastal defense became unclear. Service leaders argued over who should conduct reconnaissance and strike from U.S. bases, and which, if any, aircraft would attack enemy ships near shore. General Billy Mitchell, the leading proponent of expanding the roles and missions of the Army Air Corps, argued that land-based bombers were best suited for attacking enemy ships, while the Navy claimed it should have sole authority for this mission.[203] Mitchell hoped to settle the question in his favor in 1921 when he bombed of a fleet of obsolete, captured German warships, including a submarine, destroyer, cruiser, and the "unsinkable" battleship *Ostfriesland*. He felt he had proved the Air Corps should have the coastal defense mission, despite the

---

[200] Spector, 74.
[201] Dupuy and Dupuy, 1124-1125, 1142, 1145.
[202] Spector, 81-82.
[203] Spector, 86-87.

artificial conditions of the test and ignoring the defensive capability of antiaircraft artillery.[204] However, the Navy did not agree, and bickering over the coastal defense mission continued through the 1930's, with no service obtaining exclusive jurisdiction over coastal defense.[205]

Billy Mitchell's Analysis of U.S. Strategic Vulnerability

But Mitchell's comprehension of the strategic capability of aircraft went even further. Ever-increasing aircraft capability had, in essence, "shrunk" the world. The oceans were no longer a hedge against attack on our coasts. He recognized the fact that aircraft, not ships, were now the greatest threat to both coastal and interior cities, rendering both the United States and our overseas possessions vulnerable to enemy attack. He was most concerned about the vulnerability of our outlying Pacific possessions (Philippines, Guam, and Hawaii) to Japanese naval and air power.[206] He went so far as to predict in 1924 that if Japan were to start a war with the U.S., Japan would strike just after dawn against key military targets in Hawaii, using carrier-based aircraft. But the Coolidge administration, obsessed with fiscal frugality, ignored his warnings, and the War Department filed his report without taking any action.[207]

The Court-Martial of Billy Mitchell

Furious that his reports were ignored, and that the Air Service did not properly maintain current aircraft and made little progress in development of new military aircraft, Mitchell lashed out at the Navy and War Departments, accusing them of "criminal negligence" and "almost treasonable administration of our national defense" in the

---

[204] Weigley, 227-228.
[205] Spector, 86-87.
[206] Weigley, 228-230.
[207] Weigley, 230.

aftermath of several fatal Naval aircraft crashes in September 1925. He was court-martialed for his remarks, but continued to publicly press his views on the strategic importance of airpower until his death in 1936. He advocated Guilio Douhet's theories that air power could win wars alone by attacking the enemy's vital centers of population and production, destroying the enemy's will to resist. Although Mitchell was eventually proved incorrect in his belief that strategic bombing alone could win wars, he was one of the first to realize the terrible capabilities of airpower and the fact that it made traditional means of homeland defense obsolete.[208]

The Beginning of Strategic Airpower Doctrine

The Air Corps Tactical School (ACTS) took Mitchell's and other great air power theorists' ideas and used them to develop a strategic airpower doctrine in 1932 which emphasized destruction of vital industrial, economic and social structures which supported both the enemy's war effort as well as civilian life.[209] Clearly, this was a doctrine meant to use strategic bombers in offensive action against a foreign enemy. But the War Department refused to allow the ACTS to teach or plan air campaigns against foreign territories—only analysis of our own defense requirements was allowed, in accordance with national military policy. The ACTS realized, however, that analyzing critical target systems within the U.S. would serve as a model for critical target systems in any industrialized country—including Germany and Japan.[210] Furthermore, should a

---
[208] Weigley, 233-241.
[209] Haywood S. Hansell, Jr., *The Strategic Air War Against Germany and Japan: A Memoir* (Washington, D.C.: Government Printing Office, 1986), 6-11.
[210] Hansell, 11-12.

hostile power attack the U.S., the Panama Canal, or other possessions in the Western Hemisphere, clearly airpower would be a primary means of defense.[211]

Although the ACTS built the foundation of doctrine for employment of strategic airpower, they could not implement it, because the Army Air Corps had none of the resources required to execute the doctrine in the early 1930's.[212] Only when the Army bought the first 13 B-17 four-engine bombers in 1936 did the doctrine begin to become a capability.[213] In response to events in Europe and the Pacific in 1938, Roosevelt ordered increased production of B-17s; by December 1941, 300 were delivered.[214]

Roosevelt and Homeland Defense Policy: Hemispheric Defense

By the late 1930's, the Roosevelt administration realized that another world conflict was inevitable, and took immediate steps to improve homeland defense capability. Roosevelt changed his limited policy of defending only the United States and its possessions to a policy of *hemispheric defense*. This change was significant, because instead of strictly defensive planning, Roosevelt's new policy clearly required planning for long-range offensive operations, as well.[215]

The RAINBOW Plans

Since a two-front war was now the most likely possibility, the Joint Planning Board developed a new series of plans called "RAINBOW" plans in accordance with Roosevelt's new defense policy. RAINBOW 1 was a plan for defending U.S. territory and possessions and vital interests throughout the Western Hemisphere; RAINBOW 4

---

[211] Hansell, 26.
[212] Hansell, 16.
[213] Weigley, 241.
[214] John H. Bradley and Jack W. Dice, "The Second World War: Asia and The Pacific," in *The West Point Military History Series*, ed. Thomas E. Griess (Wayne, NJ: Avery Publishing Group, Inc., 1984), 30.

was a more aggressive version that included sending forces to South America and the Eastern Atlantic. RAINBOW 2 and 3 were plans for a Pacific-oriented strategy in a two-ocean war. RAINBOW 5 emphasized operations in Europe, the Atlantic, and Africa, with a defensive strategy against Japan until the threat across the Atlantic was eliminated.[216] Since the heart of America's industry and government lay on the Atlantic coast, and since Germany appeared to be the most formidable opponent,[217] the Joint Army and Navy Planning Board, as well as the service Secretaries, approved RAINBOW 5 in May 1941.[218]

Prelude to World War II: Efforts to Increase Readiness

Concurrently, events in Europe ensured America started getting ready for war long before Pearl Harbor. Although Roosevelt proclaimed neutrality when the war began in Europe in 1939, he immediately authorized increases in Regular Army and National Guard strength to total over 460,000. He allowed production and sales of munitions to Britain and France, which helped our industry prepare for production levels required when we entered the war. By 1940, the Army ensured its Regular troops were fully equipped and engaged 70,000 men in the first corps-level maneuvers in U.S. Army history. In 1940, Congress approved the first peacetime draft of civilians into the Army in history, funded expansion of the Army to 1,200,000 men and provided enough funds to procure equipment and munitions for every soldier. The Navy quickly entered an expansion program to develop a Navy that could deal with both the Japanese and the

---

[215] Office of the Chief of Military History, United States Army, "Chapter 19, Between World Wars," in *American Military History* (Washington, D.C.: Office of the Chief of Military History, 1988), 418; on-line, Internet, 26 October 2000, available from http://www.army.mil/cmh-pg/books/amh/amh-19.htm.
[216] Weigley, 313-314.
[217] Weigley, 314.
[218] Hansell, 29.

German navies. The urgency of American preparations was based on the fast succession of German victories in the Low Countries and France, which made many planners think that America might have to face Germany alone.[219]

The Lend-Lease Act

America's facade of neutrality ended with the Lend-Lease Act in 1941. This aid program was a key part of our homeland defense strategy, for it provided aid to nations whose defense Roosevelt considered vital to our own defense[220] and bought time for the U.S. to continue its own mobilization for war.[221]

World War II

Background

The problem with U.S. defense preparations was that planners were so fixated on events in Europe, not enough attention was paid to Japanese intentions. Japan had become a militant, imperialistic state in the 1930's. Right-wing elements fanned nationalistic attitudes. The government wanted new territory to exploit natural resources and make room for its population. Buoyed by public support for imperialistic actions and a weak government, the Japanese Army invaded Manchuria in 1931 over an alleged provocation. The Japanese government acquiesced to the action, fearful that the outside world would find out it could not control its own military. By 1940, Japan joined with the Axis powers and became the equivalent of Germany in Asia. Germany's attack on Russia in 1941 gave Japan the opening it needed to proceed with the conquest of

---

[219] Office of the Chief of Military History, United States Army, "Chapter 19, Between World Wars," 418-419.
[220] Dupuy and Dupuy, 1172.
[221] Office of the Chief of Military History, United States Army, "Chapter 19, Between World Wars," 420.

Southeast Asia. With Russia's attention directed at Germany, the Japanese did not have to worry about a Russian invasion. In July 1941, Japan invaded Indochina.[222]

In response, Roosevelt froze Japanese assets in July 1941 and stopped oil shipments to Japan. Essentially, Roosevelt had placed an economic blockade on Japan. Japan could not continue its conquest of Asia without financial or oil resources. The Dutch and British, fearful for their own possessions in Asia, solidly backed the U.S. The Japanese government was boxed in when their own military demanded that their government either negotiate a settlement favorable to the Japanese, or step down and let the military take over the government. In October 1941, the civilian government stepped down; General Tojo, the Minister of War, took over and prepared for war. When the U.S. refused to back down on its demand that Japan recognize and respect the sovereignty of all nations (essentially a demand that Japan withdraw from China and Indochina), Japan decided to attack Pearl Harbor and other key Pacific locations while still maintaining a façade of negotiation.[223]

Pearl Harbor: Homeland Defense Failure

The fact that America was taken by surprise on December 7, 1941, is incredible. Every war plan in the ORANGE series had assumed that war with Japan would begin with a surprise attack, and Pearl Harbor was often mentioned as a likely primary target.[224] Billy Mitchell had prophesied years earlier that the Japanese would attack Pearl Harbor and Schofield Barracks using carrier-based aircraft if a war began in the Pacific.[225] The U.S. had broken the Japanese radio code, and based on intercepted transmissions, had

---

[222] Bradley and Dice, 4-5.
[223] Bradley and Dice, 5-7.
[224] Bradley and Dice, 27.
[225] Weigley, 230.

every reason to believe a Japanese attack was imminent. Yet the Japanese achieved complete strategic surprise when they attacked Pearl Harbor. The Japanese attack was a tactical success, for it ravaged the U.S. Pacific fleet. But strategically, it was a mistake. The attack aroused the fury of the U.S. and filled the people and government with a desire for vengeance. There would be no negotiated peace after Pearl Harbor, and Japan did not have the resources to exhaust America into negotiation with a protracted war.[226]

The U.S. Navy: Key to Pacific Victory

If our three aircraft carriers (*Lexington, Saratoga, Enterprise*) had not been out of Pearl Harbor at the time of the attack, America might not have been able to prevent Japanese victory in the Pacific. Admiral Nimitz, the new Pacific Fleet Commander, ordered the carriers to harass the Japanese, buying time for the U.S. to rebuild its fleet. Although the *Saratoga* was severely damaged by a Japanese torpedo in January 1942, the *Yorktown* quickly arrived as a replacement.[227]

These carriers literally saved the U.S. in the first year of the Pacific war.[228] In the Battle of the Coral Sea in May 1942, they prevented the Japanese from taking Port Moresby, New Guinea. Although the battle was at best a draw, this first great carrier battle prevented the Japanese from completing a strategically vital offensive thrust in the South Pacific.[229] Without Port Moresby under their control, the Japanese could not control critical sea lines of communication between Australia and the strategically

---

[226] Bradley and Dice, 17.
[227] Dupuy and Dupuy, 1232-1235.
[228] Dupuy and Dupuy, 1251.
[229] Bradley and Dice, 104-108.

important islands of New Guinea, the Solomon Islands, and the Southern Resources Area (comprising much of the current nations of Indonesia and Malaysia).[230]

This battle set the stage for the most strategically significant battle of the Pacific—Midway. Erroneously believing that (a) the U.S. carrier fleet had been destroyed in the Battle of the Coral Sea, and (b) that the Americans had fallen for a feint in the Aleutians as the primary attack, the Japanese carrier fleet attacked Midway Island on June 4, 1942. But the American carriers were waiting, and in one of the most decisive battles in history, destroyed the Japanese carrier fleet.[231]

From a homeland defense perspective, the importance of this battle cannot be overstated. With this victory, the Americans took the initiative from the Japanese, and from then on the Japanese were on the defensive.[232] But most importantly this battle saved the Germany-first strategy of RAINBOW 5, allowing the U.S. to commit the necessary resources for victory in Europe.[233]

The Battle of the Atlantic

The American homeland defense effort in the European theater did not begin auspiciously, either. America was highly dependent on imports of many basic resources (including oil) to maintain its economic and industrial infrastructure. Furthermore, by 1942, the Allies depended heavily on American-made weapons and munitions, most of which were transported by sea. But despite over 3 years of preparatory time, by

---

[230] Edward J. Krasnoborski and George Giddings, "Atlas for The Second World War: Asia and the Pacific" in *The West Point Military History Series*, ed. Thomas E. Griess (Wayne, NJ: Avery Publishing Group, Inc., 1984), Map 4.
[231] Dupuy and Dupuy, 1253-1256.
[232] Dupuy and Dupuy, 1256.
[233] Bradley and Dice, 115.

December 1941 the U.S. Navy had failed to acquire enough armed escort ships and aircraft or develop a strategy to protect our merchant vessels and oil tankers.[234]

Germany's submarine fleet commander, Admiral Karl Doenitz, realized this fact before the Americans did. His submarine warfare campaign against unprotected merchant vessels and tankers was a masterpiece—and a massacre. Between January and April 1942, Germany sank 87 merchant ships, often in broad daylight and in view of Americans on the coast. The worst possible scenario had occurred: the U.S. Navy no longer controlled its sea lines of communication either in the Atlantic or even on our own coast.[235]

America had neither the resources nor the organization to wage an effective anti-submarine warfare (ASW) campaign. Over a decade of squabbling between the Army and Navy over whose planes would have which missions had left the U.S. without enough aircraft and trained pilots to provide effective ASW patrols for critical Atlantic and Gulf of Mexico sea lines of communication.[236] Furthermore, the Navy did not have enough ASW escorts in 1942, thanks to bureaucratic laziness and the Navy's failure to agree on the design of an ASW escort. Only after the German submarine campaign began did the Navy finally begin a program to build small destroyer-type escorts to protect shipping.[237]

The war in the Atlantic finally turned when the U.S. built enough ships and aircraft to defeat Doenitz' forces. Coupled with the Allied breaking of the German submarine communications codes, and changing of the Allies' own convoy codes, the German

---

[234] Thomas B. Buell et al., "The Second World War: Europe and the Mediterranean," in *The West Point Military History Series*, ed. Thomas E. Griess (Wayne, NJ: Avery Publishing Group, Inc., 1984), 215-219.
[235] Buell et al., 209-211, 215-219.
[236] Buell et al., 218-219.

dominance of the Atlantic was over. Convoys of armed escort ships with the latest radar and weapons, land-based ASW aircraft, and locations of German submarines obtained from decoded transmissions, savaged Doentiz' submarines, and he was forced to withdraw his forces to the central Atlantic. By July 1943, production exceeded losses to German submarines.[238]

Victory in the Atlantic

The American victory in the Atlantic was absolutely critical to winning the war in Europe. Although Germany had to be defeated by a land war, the Allies had to gain control of the Atlantic to do it. For the 12 months preceding the D-Day invasion, merchant ships safely passed through the Atlantic, carrying essential supplies and men for the great invasion.[239]

The victory in the Atlantic also ended the threat to the American homeland from Germany. Without control of the Atlantic, Germany could not hope to threaten America's east coast any longer.

Strategy of Annihilation

After the victories in the Battle of the Atlantic and Midway ended any viable threat to the continental U.S. and its nearest territories, America and its Allies pursued a strategy of annihilation in the tradition of General Grant and General Sherman in the U.S. Civil War: total war, aimed at destroying the enemy's economic, industrial, and military capability, sparing neither civilians nor military targets.[240] The strategic bombing

---

[237] Buell et al., 216-219.
[238] Buell et al., 222-225.
[239] Buell et al., 224-225.
[240] Weigley, 151-152, 357-359, 363-365.

campaigns in both Europe and Japan pursued the same goal: to bring about collapse of the enemy state and destroy its war-making capacity.[241]

The effectiveness of the strategic bombing campaign on Germany's industrial capacity and civilian morale has been argued for many years. But Albert Speer, Hitler's Minister of Armaments and Munitions, and the authors of the post-war U.S. Strategic Bombing Survey concluded that the strategic bombing campaign caused the collapse of the German armaments industry.[242] Furthermore, the strategic bombing campaign ended Germany's air superiority over the European continent. By D-Day, June 6, 1944, Germany could not provide air support to its own forces or challenge the Allied invasion force. The strategic bombing campaign was crucial to the success of the Allied land invasion of Europe.[243]

In Japan, the strategy of annihilation through strategic bombing went past even that seen in Germany. Previous to December 1944, the Joint Chiefs placed priority on selectively targeting elements of Japan's war-making capacity: shipping, aircraft factories, and heavy industry. But constant bad weather, and the need to complete strategic bombing objectives prior to the proposed November 1945 invasion, caused the U.S. to adopt a far more destructive, morally questionable strategy.[244] In December 1944, the Joint Chiefs changed the priority to incendiary bombing of urban areas (while still bombing selected military and industrial targets) to completely destroy Japan's war-making capacity. Cities were crucial to Japan's war effort, since many small factories

---

[241] Hansell, 79, 216-217, 228
[242] Weigley, 358, and Hansell, 128-129.
[243] Hansell, 119.
[244] Hansell, 175-176, 217-218, 263-264

within urban centers supported the war effort.[245] The strategic bombing campaign destroyed Japan's economy and severely damaged its war-making capacity.[246] However, despite the fact of defeat, Japan's military leadership refused Truman's demand of unconditional surrender. Faced with the prospect of enormous American casualties in an invasion of Japan, Truman decided to use atomic bombs to force Japan's capitulation.[247]

The Atomic Bomb

The atomic bomb's effect on American defense strategy was revolutionary. For a short time, America had the sole power to destroy any nation; the atomic bomb literally negated conventional threats to America's territory. But as soon as the Russians acquired the atomic bomb, American strategy had to place a premium on deterrence. For now, it, too, faced the prospect of annihilation.[248]

Analysis of American Homeland Defense: The Expansionist era Through World War II

During the 80 years between the end of the Civil War and the end of World War II, successive administrations made many of the same mistakes as their predecessors, despite America's new role as an industrial and colonial power:

    a.   America's acquisition of new states and territories required the military power to defend them. However, Congress continually failed to provide the necessary funds for an adequate standing military force to support America's expansionist policies. With a few exceptions (such as Secretaries of the Navy Herbert and Tracy), civilian policy makers

---

[245] Hansell, 216-220, 226-228.
[246] Hansell, 247-249.
[247] Bradley and Dice, 254-259.

continually failed to comprehend the need for military resources to match defense policies. After the Spanish-American War, the Army had only 75,000 men to defend America's new territories plus the continental United States. During the Wilson era, isolationist attitudes among the public and administration ensured little attention to military planning or requirements. When the U.S. entered World War I, neither the Army nor the National Guard was ready for deployment. The 1920s and 1930s found the military in the same predicament—Congress severely cut funding for both the Army and Navy forces. Congress entered into international agreements in 1922 and 1930 that so limited the Navy's fleet composition and basing authorizations, that the Navy could not execute the approved war plan to defend America's Pacific territories. Only when Roosevelt realized that America's entry into World War II war was inevitable, did he take action to enhance military readiness, and his efforts were almost too late. Failure to build enough ships or aircraft to protect merchant shipping from German submarines temporarily cost the U.S. control over its sea lines of communication in the Atlantic.

b. States still did not provide militias the funds or training to maintain any degree of military proficiency. As a result, militias were completely unprepared to go to war in 1898. The Dick Act of 1903 finally set federal requirements for funding and training of state militias in an attempt to provide a better trained and equipped volunteer force.

---

[248] Weigley, 365.

c. Lack of coordination between Congress and the War Department resulted in no coordinated planning for mobilization, strategy for force employment, or coordination of military strategy with policy objectives. The effect of poor prior planning was particularly noticeable in the Spanish-American War, especially in the Army. Mobilization and transport of troops and supplies were needlessly held up, no real plan existed for conduct of the war, political interference from Washington prevented the Navy from enforcing a successful blockade of Cuba, and there was no coordination between Navy and Army forces to complete military objectives. The same problems occurred in mobilizing for World War I, on a greater scale. Without adequate planning for transportation of supplies for the war effort, ports became clogged and soldiers did not receive critical supplies (such as rifles and blankets) in a timely manner. Both World War I and the Spanish-American War highlighted the need for long-term, centrally controlled and coordinated planning, production, and transportation of military equipment to effectively support homeland defense needs, especially at the onset of a national crisis.

d. Not until 1903 did the U.S. take action on strategic defense planning. A Joint Army and Navy Board developed the first plans for hemispheric defense and defense of U.S. possessions (War Plans ORANGE and BLACK), but the plans were unrealistic, given America's limited military resources. Furthermore, in the late 1930's, America's fixation on the German threat resulted in inattention to planning to meet the growing

Japanese threat, culminating in the disaster at Pearl Harbor. But America's entry as a belligerent in World War II gave impetus to development of sound, joint strategic planning which ensured victory in both the Pacific and European theaters.

America did, however, make some improvements to its homeland defense capability:

a. The Army and Navy each developed programs devoted to professional education of military officers. These new institutions emphasized the study of military history and strategy—efforts that paid enormous dividends in enhancing the quality of professional officers and strategic planning during both World War I and World War II.

b. A few civilian policy makers realized the need to coordinate foreign policy objectives with military capability. Navy Secretaries Herbert and Tracy successfully pushed for construction of more battleships just prior to the Spanish-American War—saving the American war effort. Similarly, when America entered World War I, the Navy had enough resources to assist the Allies from the start in protecting shipping from U-boats.

c. The National Defense Act of 1916 was significant in the history of homeland defense, for it marked a realization that national defense was not solely the purview of the military and the government—it required the support of the population and national industry. It gave the President authority to call on state militias for national defense, and it empowered the President to gear up the U.S. economy to support a major war effort.

d. To prevent the same logistical problems encountered in mobilizing for the Spanish-American War and World War I, the National Defense Act of 1920 centralized procurement planning to streamline supply and purchasing procedures, and in 1924, the Army began training officers on the principles of sound logistics.

e. The ACTS developed a strategic airpower doctrine that was the basis for the winning strategies of strategic bombardment in both Europe and the Pacific in World War II, and set a precedent for development of strategic airpower planning and doctrine for the future.

Chapter 4

American Homeland Defense: The Cold War

Little minds try to defend everything at once, but sensible people look at the main point only...If you try to hold everything, you hold nothing.

—Frederick the Great

Truman: Containment

Cold War: A Revolution in Homeland Defense

The end of World War II forever ended America's isolationist stance in world affairs. Airpower and atomic weapons negated the former protection the oceans provided,[249] and America would never again have the luxury of months to mobilize its forces (as in World War I and World War II) when faced with a powerful aggressor.[250] Furthermore, the U.S. and Soviet Union emerged as the two most powerful nations in the world, but the relationship between the Soviet Union and the U.S. rapidly deteriorated after World War II into a "Cold War" as the U.S. and Soviet Union vied for global power and influence.[251]

This new security environment forever changed the U.S. concept of homeland defense. Now, homeland defense encompassed more than the physical security of U.S. territory, population, and government. The health of the U.S. economy, access to vital

---

[249] Office of the Chief of Military History, "Chapter 24, Peace Becomes Cold War, 1945-1950," in *American Military History* (Washington, D.C.: Office of the Chief of Military History, 1988), 529; on-line, Internet, 30 November 2000, available from http://www.army.mil/cmh-pg/books/amh/amh-24.htm.
[250] Roy K. Flint, Peter W. Kozumplik, and Thomas J. Waraksa, "The Arab-Israeli Wars, The Chinese Civil War, and The Korean War," in *The West Point Military History Series*, ed. Thomas E. Griess (Wayne, NJ: Avery Publishing Group, Inc., 1987), 73.

resources, control of global lines of communication, and containment of the new international political-military entity of communism became essential aspects of U.S. homeland defense. Homeland defense was now inextricably tied to the strategic balance of power between the U.S. and Soviet Union. As such, the relative strength and weakness of U.S. and Soviet nuclear forces, conventional forces, economic power, political influence, and later, counter-terrorism policy directly affected U.S. ability to defend its territory, populace, and vital interests.

To enhance their own global power and influence, the Soviets (and eventually communist China) used a strategy of exploiting people's grievances against their governments (poverty, lack of basic services and rights, government brutality) to promote communist revolution in Asia, Latin America, and Africa. Hence, containment of Soviet expansionism became the primary objective of U.S. security policy.[252]

Collective Security

America now had to take the lead in promoting democracy and containing communist expansion to maintain an effective homeland defense. But it could not act alone to achieve its security goals. Now the U.S. had to enter into permanent international agreements and organizations for collective security.[253]

The first such collective security organization which America joined and helped found was the United Nations, a body initiated for employment of "measures for the prevention and removal of threats to the peace and for the suppression of acts of aggression…"[254] In this spirit, in 1947, the Truman administration also developed the

---

[251] Dupuy and Dupuy, 1310-1312.
[252] Dupuy and Dupuy, 1310-1312.
[253] Weigley, 366.
[254] Office of the Chief of Military History, "Chapter 24, Peace Becomes Cold War, 1945-1950," 529.

first regional collective defense treaty under United Nations auspices—the Inter-American Treaty of Reciprocal Assistance (Rio Treaty). Signed by 21 republics in the Western Hemisphere, the treaty basically stipulated that an attack on one country would be considered an attack on all. In addition, the 1948 Brussels Treaty between Great Britain, France, Belgium, the Netherlands, Luxembourg, and the U.S., was soon followed by the development of the 1949 North Atlantic Treaty Organization (NATO)—another European collective security arrangement designed to counter the growing Soviet threat in Eastern Europe.[255] This was America's first peacetime military alliance with foreign states in which U.S. forces were permanently stationed in foreign countries outside a major theater war.[256] Through the NATO alliance, the U.S. provided advanced arms preferentially to NATO members. Additionally, NATO was a key element of U.S. foreign policy and influence through establishment of U.S. bases in member countries, storage and deployment of nuclear weapons, and re-armament of West Germany. The strength of the NATO alliance became the key security barrier to Soviet expansionism in Europe and a linchpin for U.S. homeland defense.[257]

National Security Act of 1947

Congress also passed domestic legislation in an attempt to improve defense capability. In an attempt to correct deficiencies in organization and cooperation in the national security policy apparatus, Congress passed the National Security Act of 1947. This was an attempt to provide unified command and control of military and civilian

---

[255] Office of the Chief of Military History, "Chapter 24, Peace Becomes Cold War, 1945-1950," 533-538, 542-543..

[256] Amos A. Jordan, William J. Taylor, Jr., and Lawrence J. Korb, *American National Security: Policy and Process*, 3rd ed. (Baltimore, MD: Johns Hopkins University Press, 1989), 64.

[257] U.S. Army Tank-automotive and Armaments Command (TACOM) Security Assistance Center, *The Beginnings of NATO*, 1-2; on-line, Internet, 22 February 2001, available from http://www-acala1.ria.army.mil/tsac/nato.htm.

national security policy at the national level. Among its most important features, it created the National Security Council (which included the Secretary of State, the Secretary of Defense, the military service secretaries, and other government organization representatives). Its mission was to develop coordinated national security plans and policy for Presidential approval. The Act also created the National Military Establishment (which included each of the Service secretaries and the Secretary of Defense) and required the Joint Chiefs of Staff of develop plans and unified commands for key strategic areas. The Act also clarified and delineated roles and missions of the services. But the Secretary of Defense had little real authority under this Act, so in 1949, the act was amended to make the Secretary of Defense the true figure of authority in defense planning and policy.[258]

Military Professional Education

Military professional education also improved. Each service now had courses tailored to specific missions, and for the first time, three new joint schools--the Armed Forces Staff College, National War College, and Industrial College of the Armed Forces--opened to train officers respectively in joint operations, joint mobilization, and national policy.[259] These schools became an integral part of homeland defense, for they brought together military members from all services to learn how to effectively coordinate their efforts to prosecute conflicts.

The Truman Doctrine

Regarding U.S. international policy, America's monopoly on nuclear weapons drove Truman's belief that the U.S.S.R. would not challenge America directly militarily, but

---

[258] Office of the Chief of Military History, "Chapter 24, Peace Becomes Cold War, 1945-1950," 531-533.
[259] Office of the Chief of Military History, "Chapter 24, Peace Becomes Cold War, 1945-1950," 533.

would instead continue to exploit internal grievances and revolutionary movements to expand its sphere of influence, threatening U.S. homeland security by instigating political instability or establishing outright control over strategically vital countries and lines of communication. From this belief came the Truman Doctrine, a policy of providing economic and other forms of aid to alleviate conditions in overseas countries that the Soviets might exploit to their advantage.[260]

Truman's policy resulted in a massive infusion of economic aid and military assistance to Greece and Turkey starting in 1947. His policy succeeded in preventing a communist takeover of the Greek government, and prevented Stalin from forcing the Turkish government to "share" control of the strategically important Dardanelles Straits and cede Turkish territory to Soviet Georgia.[261] The Truman Doctrine is one of the most important policies in U.S. homeland defense history, for it permanently ended our isolationist view of homeland defense, and set a precedent, for good or bad, of U.S. intervention (ranging from economic assistance to war) in foreign countries to protect U.S. interests.

The Marshall Plan

The Marshall Plan also set a precedent for homeland defense. At the end of World War II, much of Europe lay in ruins. World War II devastated the economic, industrial, and transportation infrastructure in many European countries. In 1947, Truman and his Secretary of State, George Marshall, developed a plan for the economic and industrial reconstruction of most of Europe. The United States provided short-term economic aid

---

[260] Office of the Chief of Military History, "Chapter 24, Peace Becomes Cold War, 1945-1950," 530-531, 536-540.

and helped the European nations develop and maintain their own economic recovery program. Although the Soviet Union would not permit countries under its control to participate, and Spain's fascist government was not invited to participate, 16 European countries benefited from this superb plan. The plan's objectives included reconstruction of industrial, and transportation infrastructure, creation of strong currencies and national budgets, full employment, and promoting trade by ending restrictions and tariffs.[262] The Marshall Plan was a great success—by 1950, industrial production was 15 percent above prewar levels.[263]

The importance of the Marshall Plan to U.S. defense in the Cold War cannot be overstated. It prevented an international economic crisis, ensured both the economic and political independence of Western Europe, and built a permanent foundation for friendship and cooperation among the free countries of Europe. It set up lasting political, economic, and trade relationships between Europe and the U.S.[264] Most importantly, the Marshall Plan prevented the Soviet Union from taking advantage of Western Europe's weakness in the aftermath of World War II—literally, it saved Europe.

Military Readiness Declines

As successful as the Truman Doctrine was in fighting Soviet expansionism and the Marshall Plan was in supporting Western Europe's economic and industrial recovery,

---

[261] Truman Library, National Archives and Records Administration, *Harry Truman and the Truman Doctrine,* n.p.; on-line, Internet, 10 February 2001, available from http://www.trumanlibrary.org/teacher/doctrine.htm.
[262] Library of Congress, "A Survey of the Marshall Plan and its Consequences," *Marshall Plan Exhibit: For European Recovery: The Fiftieth Anniversary of the Marshall Plan,* 40-43; on-line, Internet, 10 February 2001, available from http://lcweb.loc.gov/exhibits/marshall/m41.html.
[263] Library of Congress, "The Marshall Plan Countries," *Marshall Plan Exhibit: For European Recovery: the Fiftieth Anniversary of the Marshall Plan,* 1; on-line, Internet, 10 February 2001, available from http://lcweb.loc.gov/exhibits/marshall/mars5.html.
[264] Library of Congress, "A Survey of the Marshall Plan and its Consequences," *Marshall Plan Exhibit: For European Recovery: The Fiftieth Anniversary of the Marshall Plan,* 48.

Truman still needed a strong military force to support them.[265] But, as so often in the past, the government did not give the military the resources it needed to support government policy. At the end of World War II, the U.S. Armed Forces had about 12 million people manning over 100 combat Army divisions, 1200 Navy combatant ships, and over 200 Army Air Corps strategic and tactical combat groups.[266] But public and Congressional pressure, coupled with Truman's determination to balance the national budget, ensured quick, poorly programmed demobilization immediately after World War II. As a result, the Army could not maintain its equipment, and units became non-mission capable.[267]

Secretary of Defense Louis A. Johnson wanted cheap, economical defense, based mainly on strategic air power. The Air Force placed about a third of its resources into Strategic Air Command—the command that would deliver atomic bombs in the event of nuclear war.[268] But the rest of the services, including the Air Force's non-strategic forces, suffered cuts as defense budgets dropped consistently throughout the Truman administration. He cancelled the Navy's new strategic bomber platform, the "supercarrier," precipitating the "revolt of the admirals," in which several famous admirals sent a letter to the Secretary of the Navy stating that the Navy, and the nation as well, were being stripped of offensive power.[269] Johnson resorted to the same old policy

---

[265] Office of the Chief of Military History, "Chapter 24, Peace Becomes Cold War, 1945-1950," 530-531, 536-540.
[266] Laurence J. Korb, "The Defense Policy of the United States," in *The Defense Policies of Nations: A Comparative Study*, ed. Douglas J. Murray and Paul R. Viotti (Baltimore, MD: Johns Hopkins University Press, 1982), 51.
[267] Office of the Chief of Military History, "Chapter 24, Peace Becomes Cold War, 1945-1950," 530-531, 536-540.
[268] Office of the Chief of Military History, "Chapter 24, Peace Becomes Cold War, 1945-1950," 538-540.
[269] Weigley, 371, 373, 376-377.

of holding military budgets and force structure hostage to budget ceilings, rather than real security requirements.[270]

Furthermore, military force structure and strategy had not kept up with changes in the international security environment. The armed forces insisted on preparing our limited forces to mobilize for a major war, rather than developing strategy and doctrine for fighting limited wars to contain communist expansion.[271]

America's refusal to maintain a strong military force also rested on its monopoly on nuclear weapons. As the only possessor of the atomic bomb, the American government felt no country would dare challenge it militarily.[272] Here lay the basic conflict in Truman's defense policy. Truman wanted to contain Soviet expansion, but he did not have an adequate conventional military force, and our atomic weapons capability, as yet unmatched, was still too small to be an adequate means of deterring aggression. Furthermore, our atomic bomb was of little use in suppressing guerrilla movements and smaller conflicts. Essentially, the U.S. lacked a well-reasoned doctrine for employment of atomic weapons or conventional forces.[273]

Truman and Civil Defense

Truman's defense policy shortcomings were further highlighted in the summer of 1949 when the Soviet Union exploded an atomic bomb over Siberia. No longer the only possessor of nuclear weapons, the U.S. faced a greater threat than ever before—potential

---

[270] Office of the Chief of Military History, "Chapter 24, Peace Becomes Cold War, 1945-1950," 538-540.
[271] Jordan, Taylor, and Korb, 61-62.
[272] Office of the Chief of Military History, "Chapter 24, Peace Becomes Cold War, 1945-1950," 530-531, 539-540.
[273] Jordan, Taylor, and Korb, 60-61.

destruction, despite its industry, populace, and the size of its military forces.[274] In response to the new threat, Truman established the Federal Civil Defense Administration in 1949 and Congress passed the Federal Civil Defense Act of 1950 in an attempt to develop a system of fallout shelters nation-wide. But government support for civil defense was stillborn. Truman requested over $400 million for a fallout shelter program, but Congressional priority went to tightening the national budget and Congress allocated little over $30 million for the program.[275]

NSC 68

President Truman, worried by both the communist takeover of China and the Soviet Union's new atomic capability, ordered a review of U.S. strategy and policy.[276] The result was *NSC 68: United States Objectives and Programs for National Security*. The report, issued in April 1950, recommended a concerted effort to erode Soviet influence and modify Soviet expansionist behavior through a "rapid and concerted build-up" of both U.S. and other free nations' military, political, and economic strength. It specifically recommended building military readiness "as a deterrent to Soviet aggression," as a means of "encouragement to nations resisting Soviet political aggression," and as a "basis of immediate military commitments and rapid mobilization should war prove unavoidable."[277]

---

[274] Office of the Chief of Military History, "Chapter 24, Peace Becomes Cold War, 1945-1950," 530-531, 543-544.
[275] Disaster and Emergency Services, Yellowstone County, "History of Emergency Preparedness," 1-2; on-line, Internet, 24 September 2000, available from http://www.co.yellowstone.mt.us/des/des_history.asp
[276] Weigley, 379.
[277] National Security Council, *NSC 68: United States Objectives and Programs for National Security* (Washington, D.C.: National Security Council, 14 April, 1950), n.p.; on-line, Internet, 6 January 2001, available from http://www.fas.org/irp/offdocs/nsc-hst/nsc-68-cr.htm.

Although Truman and his advisors endorsed NSC 68's conclusions, budget limitations, instead of threat analysis, again dictated action on this document. The fiscal year 1951 defense budget was limited to $13 billion, and no one in Congress was going to vote for a defense increase in an election year.[278] But Congressional reluctance to fund the recommendations of NSC 68 was overcome by events on June 25, 1950, when North Korea invaded South Korea.[279]

The Korean War

Thanks to 5 years of steadily decreasing defense budgets after World War II, the services were hardly in any shape to prosecute a major conflict. In addition to a shortage of combatant ships, the Navy had suffered numerous personnel losses and lacked experienced crews and leaders. Many combatant ships were operating at only 2/3 authorized personnel strength, and two Marine Corps divisions had only 40 percent of the personnel required.[280] The Army was in even worse shape. Army personnel were inexperienced in combat, scattered throughout Europe, Japan, the Caribbean, and the United States, and 9 out of the Army's 10 divisions were undermanned. Furthermore, the Army had spent its budget trying to maintain some combat capability, rather than modernizing its force.[281]

The outbreak of the Korean War proved the failure of America's post-war strategic policy. America's nuclear weapons were not a deterrent to aggressor nations that lacked significant industrial and population centers to target. Furthermore, a strategic policy based on threatened use of nuclear weapons was only effective if America's adversaries

---

[278] Weigley, 381.
[279] Jordan, Taylor, and Korb, 63.
[280] Flint, Kozumplik, and Waraksa, 74.
[281] Flint, Kozumplik, and Waraksa, 74-75.

believed we would use them.[282] North Korea called our bluff, and we lost. Furthermore, Truman's policy of mobilization (rather than maintenance of a strong military force) nearly caused defeat in the early days of the Korean War due to lack of a strong, experienced military force structure.[283]

China's intervention after Inchon prevented the U.S. from attaining our original goal of unification of the entire Korean peninsula under democratic government, and allowed the communists to drag the war into a stalemate.[284] The war ended with the U.S. in a strategically worse position than before the war. Communist China emerged as a newly influential world power, and communist North Korea remained a threat to South Korea.[285] Furthermore, the world now knew that America's strategic policy of relying on the threat of using its nuclear weapons to prevent war was a sham.

The Korean War did, however, initiate some improvement in U.S. defense policy. It initiated an immediate rearmament program to provide enough forces to prosecute the war in Korea, create a more efficient military mobilization capability, and develop a more balanced force to challenge aggressor nations.[286] Congress and the American public also now realized the need for larger defense budgets to support containment of communist expansion.[287]

Massive Retaliation and The New Look

The Korean War had been long, expensive, and indecisive, in part due to the Truman administration's very public political constraints in prosecuting the war: refusal to attack

---

[282] Flint, Kozumplik, and Waraksa, 75.
[283] Dupuy and Dupuy, 1355-1356.
[284] Flint, Kozumplik, and Waraksa, 120.
[285] Flint, Kozumplik, and Waraksa, 120.
[286] Jordan, Taylor, and Korb, 63-64.
[287] Flint, Kozumplik, and Waraksa, 120.

within Chinese or Soviet territory, and refusal to consider the use of atomic weapons. In 1953, the Eisenhower administration started hinting that these constraints might be lifted. The resulting progress at the armistice talks made Eisenhower realize that American nuclear superiority could, at least for a time, be a powerful diplomatic weapon. Furthermore, the cost of maintaining a conventional military capability to match America's containment policy and commitments to Allies made deterrence based on nuclear firepower even more attractive as a defense policy.[288]

In January 1954, Secretary of State John Foster Dulles announced a new defense policy of "massive retaliation," based "primarily upon a great capacity to retaliate, instantly, by means and places of our own choosing."[289] The Eisenhower administration's new policy, known as the "New Look,"[290] was a means of reducing conventional force requirements by "changing the rules of engagement for general war."[291]

The National Security Council codified this policy into a study named NSC 162. It recommended supporting our containment policy by primarily relying on nuclear weapons and the strategic airpower to deliver them, and increased emphasis on defense of the U.S. from Soviet air attack.[292] Furthermore, the New Look initiated another reduction in conventional force capabilities. Tactical nuclear weapons replaced conventional forces overseas in Western Europe.[293] To further increase NATO capability without building up

---

[288] Paret, 739-740.
[289] Statement of John Foster Dulles, Secretary of State, in "The Evolution of Foreign Policy," *Department of State Bulletin*, 30, January 25, 1954, in *Makers of Modern Strategy From Machiavelli to the Nuclear Age*, ed. Peter Paret (Princeton, NJ: Princeton University Press, 1986), 740.
[290] Jordan, Taylor, and Korb, 66.
[291] Paret, 740-741.
[292] Jordan, Taylor, and Korb, 65-66.
[293] Jordan, Taylor, and Korb, 65-66.

U.S. forces, the U.S. convinced NATO members to build up their forces and rearm the West Germans, despite strident Communist opposition.[294]

Military leaders, predictably, were not completely supportive of the New Look. Upon his retirement, General Ridgeway, Army Chief of Staff, and his successor, General Maxwell Taylor, urged the Eisenhower administration to maintain a balanced force posture to meet both general and limited war requirements.[295]

But Eisenhower knew his "New Look" policy could not last for long, since the Soviets were quickly building their own nuclear arsenal—already they were capable of attacking America's east coast. Thus, "massive retaliation" was more a means of our exploiting short-term nuclear advantage than a long-term defense strategy.[296] But even as a short-term policy, it was a dangerous idea. As William Kaufmann explained, it put us in the difficult position of "put up or shut up." It gave us a choice of potentially escalating to nuclear war over a small conflict, or "losing face" by backing down if the Soviets or Chinese would not back down.[297]

Eisenhower and Strategic Alliances

Although nuclear weapons were key to Eisenhower's containment strategy, he did not ignore the need for political alliances. The Korean War had taught the U.S. that lack of a clear U.S. commitment to any region invited communist attempts to undermine legitimate governments. Eisenhower decided to pursue a policy of alliances to friendly nations that were adjacent to communist nations. The U.S. already had mutual defense

---

[294] Office of the Chief of Military History, "Chapter 26, The Army and The New Look," in *American Military History,* (Washington, D.C.: Office of the Chief of Military History, 1988), 574; on-line, Internet, 5 January 2001, available from http://www.army.mil/cmh-pg/books/amh/amh-26.htm.
[295] Office of the Chief of Military History, "Chapter 26, The Army and The New Look," 573-574.
[296] Paret, 741.
[297] Paret, 742-743.

treaties with NATO, Japan, the Philippines, Australia, and New Zealand. Eisenhower extended this policy by entering into a 1954 treaty with Taiwan for "joint consultation" if Taiwan were in danger of attack. Similarly, the U.S. joined the Southeast Asia Treaty Organization (SEATO) in 1954 with Australia, France, New Zealand, Pakistan, and Philippines, Thailand, and the United Kingdom that required the members to act together in the event of hostile action toward any treaty members.[298]

The Eisenhower Doctrine

The Eisenhower administration also expanded on its policy of containing Soviet aggression by broadening foreign aid to support not only U.S. allies, but also non-aligned nations. This policy was especially important in the Middle East; the vast oil resources and sea lines of communication on which so much of the world depended could not be allowed to become part of the Soviet Union's sphere of influence.[299] In the aftermath of the Soviet threat to aid Egypt during the 1956 Suez Crisis, Eisenhower pledged economic and military aid to Middle East nations that were threatened by direct or indirect communist aggression.[300] This policy became known as the Eisenhower Doctrine.[301]

Under this doctrine, Eisenhower sent Marines to Beirut in 1958 at the request of Lebanon's embattled President Shamun in the aftermath of highly contentious elections which generated full-scale revolt by pan-Arab opposition forces against Shamun's government. The U.S. role in assisting Shamun was largely symbolic—Eisenhower acted

---

[298] Jordan, Taylor, and Korb, 67.
[299] U.S. Army Tank-automotive and Armaments Command (TACOM) Security Assistance Center, *The Eisenhower Doctrine*, 1; on-line, Internet, 10 February 2001, available from http://www-acala1.ria.army.mil/tsac/eisenhwr.htm.
[300] Dwight Eisenhower, "Eisenhower Doctrine 1957," 5 January 1957, *Public Papers of the Presidents, Dwight D. \cf2 Eisenhower\cf0, 1957*, 6-16; on-line, Internet, 10 February 2001, available from http://coursesa.matrix.msu.edu/~hst306/documents/eisen.html.
[301] Office of the Chief of Military History, "Chapter 26, The Army and The New Look," 580.

to reassure allies in the region that the U.S. would act on its policy. In the ensuing civil war (in which U.S. forces did not participate), Shamun was replaced by the Commander of the Lebanese Army.[302] In this first test of expanded U.S. foreign assistance policy, the President deployed forces for a questionable purpose, and they failed to achieve any desirable objective. Unfortunately, this inappropriate use of the U.S. military established a precedent of using the armed forces in support of vague security objectives—an unfortunate practice which continued throughout the 20th century.

More Military Reductions

The Eisenhower administration's interventionist tendencies were not matched, however, by the resources required to support these policies. U.S. commitments to allies necessitated the military capability to enforce them. Unfortunately, Eisenhower continued the Truman era policy of trying to shape military force structure to meet budget limitations instead of national security commitments. For example, Eisenhower's reliance on strategic air forces to execute his defense policy necessitated severe cuts in other forces. By 1958, the Army had shrunk from its 1953 level of 20 combat divisions and 1,500,000 men to 15 divisions and less than 900,000 men to cover world-wide security commitments, a dangerous policy in light of continued communist aggression in Asia and elsewhere.[303]

The New New Look

As expected, the Soviets aggressively developed an atomic capability to match our own. The Soviet intercontinental bomber, hydrogen bomb, tactical nuclear weapons, and

---

[302] _____, "Lebanon History: The Shamun Era 1952-58," n.p.; on-line, Internet, 10 February 2001, available from http://www.lebanon.f2s.com/culture/history/independent1.htm.
[303] Office of the Chief of Military History, "Chapter 26, The Army and The New Look," 581-582.

long-range missiles quickly eroded the U.S. advantage in nuclear weapons.[304] The Soviet nuclear buildup seriously undermined the New Look policy by destroying the U.S. nuclear advantage.[305]

The U.S. had to improve its homeland defense capability to match the Soviet threat. By 1957, with the assistance of Canada, the U.S. completed a defensive perimeter of warning stations across Alaska and Canada.[306] By the mid-1950's, both the Army and Air Force developed intermediate-range (1500 miles) nuclear missiles, and the Air Force proceeded with development of intercontinental ballistic missiles (ICBMs) with a 5000-mile range. Predictably, the Army and Air Force squabbled over control on intermediate-range missiles, and in 1956, the Air Force took control of both intermediate and intercontinental missiles, although the Army was allowed to complete testing of the Jupiter intermediate-range ballistic missile.[307] But the Soviets were well on their way to matching and eventually exceeding U.S. nuclear capabilities.

The Soviet nuclear buildup undermined U.S. defense policy since our nuclear forces were no longer a deterrent to Soviet instigation of small, localized conflicts. As a result, Eisenhower had to develop a credible capability to deter limited wars. But the Eisenhower administration, faced with both domestic recession and inflation, was in no mood to increase defense spending. The result was the "New New Look," which attempted to redress our military force imbalance and deter small wars—without increasing the defense budget. But instead of building up conventional forces capability

---

[304] Office of the Chief of Military History, "Chapter 26, The Army and The New Look," 575.
[305] Jordan, Taylor, and Korb, 68.
[306] Office of the Chief of Military History, "Chapter 26, The Army and The New Look," 575-576.
[307] Office of the Chief of Military History, "Chapter 26, The Army and The New Look," 576.

to deal with small wars, Eisenhower decided to rely on tactical nuclear weapons to deter or wage small wars.[308]

This policy was fundamentally flawed. The use of tactical nuclear weapons on the battlefield would almost certainly escalate any limited conflict to a general nuclear conflict.[309] Tactical nuclear weapons basically served as a symbol of U.S. commitment to Western Europe.[310]

The Soviet launch of the first satellite, *Sputnik*, in August 1957 forced Eisenhower to review his defense policies, in light of the fact that the Soviets now had rocket thrust capability far more advanced than the U.S.[311] In essence, U.S. reliance on nuclear weapons as a deterrent to Soviet aggression could no longer withstand the reality of Soviet capabilities.

Eisenhower and Civil Defense

Soviet atomic capabilities brought about renewed interest in civil defense to protect the U.S. population. But throughout the Truman and Eisenhower administrations, Congress allocated less than half (sometimes as little as 10-20 percent) of the Federal Civil Defense Administration's budget requests. Civil defense for the average person was largely self-help (home fallout shelters).[312] Based on analysis of projected casualties and destruction from a nuclear exchange, Eisenhower concluded that civil defense

---

[308] Jordan, Taylor, and Korb, 68-69.
[309] Paret, 748.
[310] Paret, 749.
[311] Office of the Chief of Military History, "Chapter 26, The Army and The New Look," 576.
[312] Dee Garrison, *Civil Defense Portrayal of Nuclear War,* PBS interview, 1; on-line, Internet, 25 September 2000, available from http://www.pbs.org/wgbh/amex/bomb/filmmore/refernce/interview/garrison2.html.

programs could reduce loss of life only to a small degree.[313] According to noted Cold War-era physicist Herbert York, civil defense was a propaganda program designed to make the populace believe that that they could survive a nuclear exchange with the Soviet Union.[314] Instead of promoting fallout shelters, Eisenhower changed his civil defense strategy to ease mass evacuation from target cities. The 1956 Interstate Highway Act was a part of this strategy to ease traffic movement away from cities. Of course, this idea was not realistic, either, considering the limited warning time expected plus the traffic jam expected when everyone in a large city tried to leave at once.[315] Essentially, Eisenhower's civil defense efforts were no more successful than Truman's were.

Kennedy and Johnson: The Risks of Unintended Consequences

Flexible Response

When President Kennedy entered office in 1961, he was well aware of the weaknesses of the Eisenhower defense policy, and he initiated an effort to develop a more balanced force structure. Neither massive retaliation nor tactical nuclear weapons were adequate to stop insurgencies boiling in Africa, Latin America, and Asia,[316] nor to end Soviet economic and military support for revolutionary movements, especially in

---

[313] General Andrew Goodpaster, *Eisenhower's Civil Defense Program*, PBS interview, 1; on-line, Internet, 6 January 2001, available from http://www.pbs.org.wgbh/amex/bomb/filmmore/reference/interview/goodpaster02.html.
[314] Herbert York, *Civil Defense is Propaganda*, PBS interview, 1; on-line, Internet, 25 September 2000, available from http://www.pbs.org/wgbh/amex/bomb/filmmore/reference/interview/york2.html.
[315] Laura McEnaney, *America's Evacuating Cities*, PBS Interview, 1; on-line, Internet, 6 January 2001, available from http://www.pbs.org/wgbh/amex/bomb/filmmore/reference/interview/mcenaney05.html.
[316] Jordan, Taylor, and Korb, 70.

economically unstable former colonies in Africa.[317] His administration officially shifted from nuclear to conventional forces as the primary means of deterrence.[318]

Kennedy abandoned budget restraints as a basis of force sizing. Instead, services were initially allowed to develop plans and programs with little or no coordination between services. For example, the Army's force planning was based on a prolonged war of attrition, while the Air Force expected a short war with its nuclear bombers. The lack of coordination between the services did little to correct force structure problems. So the new Secretary of Defense, Robert S. McNamara, decided to centralize planning and programming of U.S. military force structure and increase combat strength immediately. He and Kennedy were determined to develop a force structure to meet any threat, conventional or nuclear. This strategy became known as "Flexible Response."[319]

Flexible Response provided a number of options, both nuclear and conventional for defense. Kennedy initiated development of a nuclear force structure that could be used for massive retaliation, or limited countervalue and counterforce attacks. His programs resulted in a nuclear buildup that the Soviets could not match until the end of the 1960's.[320]

Kennedy's Nuclear Policy

Kennedy's and McNamara's approach to use of nuclear weapons differed from previous administrations, as well. Kennedy discarded "first-strike" as U.S. nuclear policy. He emphasized building a nuclear inventory large enough to deter any attack on

---

[317] Office of the Chief of Military History, "Chapter 27, Global Pressures and the Flexible Response," in *American Military History,* (Washington, D.C.: Office of the Chief of Military History, 1988), 592; on-line, Internet, 5 January 2001, available from http://www.army.mil/cmh-pg/books/amh/amh-27.htm.
[318] Office of the Chief of Military History, "Chapter 27, Global Pressures and the Flexible Response," 592.
[319] Jordan, Taylor, and Korb, 71.
[320] Jordan, Taylor, and Korb, 71-72.

the U.S. and its allies.[321] Furthermore, in 1962, Kennedy and McNamara shifted to the "no cities" targeting policy: in event of a nuclear war, the U.S. would target enemy military forces, not civilian populations. Furthermore, he announced a nuclear force structure capable of second-strike capability.[322] He announced this policy in hopes of deterring attacks on U.S. population centers, and to convince the Soviets that any nuclear attack on our allies would result in nuclear attack on their forces, destroying their military capability.[323]

But his policy had some severe defects. A viable counterforce capability requires exact intelligence and targeting information. And even a small increase in Soviet capability would have required a much greater increase in U.S. nuclear weapons to maintain counterforce capability. And for a counterforce strike to be effective, our missiles would have to launch a preemptive first strike, rather than wait until Soviet missiles had already left their silos—otherwise, what would be the point of launching a counterforce strike against empty silos? Furthermore, there was no guarantee that U.S. nuclear forces would weather a Soviet first strike in any condition to retaliate.[324]

Kennedy and Civil Defense

At the same time that McNamara and Kennedy were developing and changing U.S. nuclear targeting strategy, Kennedy was trying to stimulate Congressional interest in civil defense. He tried to revive government support for a civil defense program in 1961[325]

---

[321] _____, "Robert S. McNamara," *SecDef Histories*, 1-2; on-line, Internet, 8 January 2001, available from http://www.defenselink.mil/specials/secdef_histories/bios/mcnamara
[322] Robert S. McNamara, Secretary of Defense, address at Ann Arbor, Michigan, June 1962; on-line, Internet, 15 September 2000, available from http://www.nuclearfiles/org/docs/1962/620709-mcnamara.html.
[323] _____, "Robert S. McNamara," *SecDef Histories*, 2-3.
[324] Weigley, 444.
[325] President John F. Kennedy, "Special Message to Congress on Urgent National Needs," 25 May 1961; on-line, Internet, 25 September 2000, available from http://www.cs.umb.jfklibrary/j052561.htm.

and in the aftermath of the Cuban Missile crisis. But after his death, Congress discontinued any federal funding and stocking of fallout shelters.[326] McNamara stated, "Defense of our cities against a Soviet attack would be a futile waste of our resources."[327] He feared that a viable civil defense system, rather than acting as a deterrent to Soviet attack, would instead have the opposite effect.[328] As a result, protection of the U.S. populace was not an important part of U.S. defense policy after the Kennedy era.

Kennedy and Conventional Forces Buildup

Kennedy also built up conventional forces to respond to different types of conventional threats. He increased the Army from 12 to 16 divisions, enlarged the Navy's surface fleet, and built up both the Reserves and National Guard. He also significantly enhanced counterinsurgency force capability in an effort to stop communist insurgencies.[329]

Failure to Develop Policy for Use of Conventional Forces

The major problem with Kennedy's policy of using counterinsurgency forces was that neither McNamara nor Kennedy ever built a useful policy for application of military forces. In his inaugural address, Kennedy stated, "Let every nation know, whether it wishes us well or ill, we shall pay any price, bear any burden, meet any hardship, support any friend, or oppose any foe to assure the survival and the success of liberty."[330] The problem with this policy was that communist insurgencies were spreading worldwide: in Laos, Vietnam, the Congo, Algeria, and several other countries in Latin America, Africa,

---

[326] _____, "Civil Defense: Evacuous," *The Economist*, 18 November 1978, 20.
[327] Robert S. McNamara, quoted by Jack Swift, "Strategic Superiority Through SCI," *Defense and Foreign Affairs* (December 1985): 17.
[328] Paret, 757.
[329] Jordan, Taylor, and Korb, 72.
[330] John F. Kennedy, Inaugural Address, January 20, 1961; on-line, Internet, 25 September 2000, available from http://www.bartleby.com/124/pres56.html.

and Asia.[331] But neither Kennedy nor McNamara provided any guidance as to what threshold had to be met to employ U.S. forces, nor which conflicts would be considered legitimate use of force in the national interest.[332]

Kennedy's Vietnam Escalation

This policy led to escalation of U.S. economic and military support for President Diem's corrupt and brutal South Vietnamese government against a communist insurgency. Kennedy was concerned that if the communists took over South Vietnam, other Southeast Asian nations would soon follow.[333]

By 1963, Kennedy increased the number of U.S. advisors in South Vietnam (including Special Forces personnel) to over 16,000. After Ambassador Henry Cabot Lodge advised Kennedy that the war could not be won if Diem remained in power, Kennedy took no action to prevent the coup and assassination of Diem, resulting in a succession of regimes that became completely dependent on the U.S. for survival. After Kennedy's death, Johnson did not want to be blamed for losing Vietnam (and thus American credibility), and U.S. combat troops soon entered Vietnam.[334]

Unintended Consequences of Vietnam Escalation

Although Kennedy meant well with his policy of rendering economic and military assistance to countries fighting insurgencies, he lacked comprehension of the long term, unintended consequences of his policy. By sending military troops to South Vietnam, he dragged the U.S. into an open-ended commitment against a determined adversary in a country of secondary importance to U.S. security interests. President Johnson made the

---

[331] Office of the Chief of Military History, "Chapter 27, Global Pressures and the Flexible Response," 593.
[332] Jordan, Taylor, and Korb, 72.
[333] The History Place, "The Vietnam War: America Commits 1961-1964," n.p.; on-line, Internet, 11 February 2001, available from http://www.historyplace.com/unitedstates/vietnam/index-1961.html.

situation much worse through both his and Secretary of Defense McNamara's bungling and micromanagement of the war. They prosecuted the war without clear, achievable objectives, and failed to develop a comprehensive, workable strategy.[335] The war eventually cost over 50,000 U.S. lives and failed to achieve its objective of preventing communist takeover of South Vietnam.

Alliance for Progress

Another foreign assistance failure of the Kennedy-Johnson years was the Alliance for Progress, a program of economic assistance to encourage economic growth in Latin America. The intent of this program was to eradicate conditions in Latin America that encouraged growth of communist insurgent movements. In other words, the program was an attempt to prevent Cuba from expanding its sphere of influence. But unlike Europe's Marshall Plan, support from Congress was uneven, and the program died by the end of the 1960's.[336] In the end, the goal of preventing Cuba from extending its influence conflicted with the Alliance for Progress's goal of promoting democracy in Latin America. For example, in Brazil and Argentina, the U.S. ended up supporting dictators (who were not interested in economic reform) to prevent them from aligning with the Soviet bloc.[337]

---

[334] The History Place, "The Vietnam War: America Commits 1961-1964," n.p.
[335] Rick Young, "Lessons of Vietnam: A Conversation with Major H.R. McMaster," PBS Online, 1999, n.p.; on-line, Internet, 11 February 2001, available from http://www.pbs.org/wgbh/pages/frontline/shows/military/etc/lessons.html.
[336] U.S. Army Tank-automotive and Armaments Command (TACOM) Security Assistance Center, *The Kennedy and Johnson Administrations,* 1; on-line, Internet, 10 February 2001, available from http://www-acala1.ria.army.mil/tsac/kenjohns.htm.
[337] Thomas E. Skidmore, Peter H. Smith, "The Rise of U.S. Influence," in *Modern Latin America*, 2nd ed. (Oxford University Press, 1989), 340-368; on-line, Internet, 10 February 2001, available from http://www.mty.itesm.mx/dch/deptos/ri/ri-802/lecturas/lecvmx024.html.

The Bay of Pigs

Kennedy initiated another foreign policy disaster when he tried to keep a foolish 1960 campaign promise[338] and initiated the April 1961 Bay of Pigs invasion in an ill-advised attempt to overthrow Fidel Castro. The Bay of Pigs invasion was poorly conceived and launched without adequate air or naval support,[339] and compelling national security need for the operation was doubtful. Furthermore, the failure emboldened Soviet Premier Khrushchev in June 1961 to demand that western troops withdraw from Berlin within 6 months, initiating the Berlin crisis.[340] The crisis caused refugees to pour out of Berlin until the communists built a wall to stop the exodus. Kennedy quickly reinforced American forces in Europe, and the crisis abated by mid-1962.[341]

The Cuban Missile Crisis

But the fallout from the Bay of Pigs disaster wasn't over—it brought about the greatest crisis in the Kennedy Presidency and the greatest threat to the U.S. homeland seen to date in the Cold War—the Cuban Missile Crisis. Khrushchev, angered over the Bay of Pigs invasion, decided to send offensive weapons into Cuba. By October 1962, the U.S. had photographic proof of Soviet intermediate-range ballistic missiles in Cuba, despite his administration's warning to the Soviets not to put them there.[342] U.S. nuclear forces went on alert, fighter-interceptor squadrons and missile battalions were moved south to improve air defenses, and the Army moved over 30,000 troops south to prepare

---

[338] Weigley, 454.
[339] Office of the Chief of Military History, "Chapter 27, Global Pressures and the Flexible Response," 593.
[340] Office of the Chief of Military History, "Chapter 27, Global Pressures and the Flexible Response," 593-594.
[341] Office of the Chief of Military History, "Chapter 27, Global Pressures and the Flexible Response," 594.
[342] Weigley, 453-454.

for an invasion.³⁴³ American aircraft and warships quarantined Cuba. The U.S. and the Soviet Union went to the brink of war. The crisis ended when Khrushchev agreed to remove the missiles from Cuba and Kennedy agreed not to invade Cuba, to lift the quarantine of Cuba, and withdraw U.S. missiles from Turkey.³⁴⁴ The Cuban Missile Crisis was clearly not a total U.S. defense policy success.

But the outcome of the Cuban Missile Crisis caused even further negative implications for U.S. homeland defense. The Soviets were determined not to suffer another international embarrassment due to lack of strategic naval power. After Khrushchev's ouster in October 1964, the Soviets began a massive buildup of naval capability. By the 1970's, Soviet naval fleets were present in every ocean in the world and threatened U.S. strategic lines of communication—America was no longer the unchallenged naval power of the world.³⁴⁵

Consequences of Failure to Develop Clear Strategy

Regarding foreign assistance as a tool in homeland defense, the Bay of Pigs, the Alliance for Progress, and the decision to escalate U.S. involvement in South Vietnam all illustrate the danger of failure to develop a rational strategy to support clear, attainable, finite objectives. Unlike the Marshall Plan and the Truman Doctrine, both of which a) had clear, finite goals, b) avoided commitment of U.S. combat troops to resolve foreign internal problems, and c) supported legitimate governments, Kennedy's forays into foreign assistance deteriorated and failed because a) he failed to understand the limits of what foreign assistance can achieve, b) he failed to consider the long-term, unintended

---

[343] Office of the Chief of Military History, "Chapter 27, Global Pressures and the Flexible Response," 594-596.
[344] Dupuy and Dupuy, 1454-1455.
[345] Weigley, 455.

consequences of his programs, c) he failed to provide the necessary level of support when he decided to commit to use of military force at the Bay of Pigs, d) his objectives in Vietnam and Latin America were vague and not necessarily achievable given the type of assistance he rendered, and e) he provided assistance to governments of highly questionable legitimacy who took advantage of U.S. support to maintain power (not always successfully). In the end, Kennedy's actions damaged homeland defense capability by eroding U.S. credibility as a supporter of democracy and bringing the U.S. to the brink of nuclear war.

Johnson and Nixon: The Downward Spiral

McNamara and "Mutual Assured Destruction"

The Presidencies of Johnson and Nixon did not undo the damage to U.S. credibility and homeland defense capability that Kennedy created. In fact, their policies further eroded U.S. strategic nuclear capability by allowing the Soviets to gain a nuclear advantage over the U.S. and destroyed the first efforts at U.S. ballistic missile defense programs.

By the mid-1960's, McNamara changed U.S. nuclear policy in response to obvious weaknesses and illogic in his counterforce strategy. He came up with a new concept called "assured destruction," which meant the ability to inflict unacceptable damage on the enemy after the enemy launched a first strike.[346] In numeric terms, it meant capability to destroy 20-25 percent of the Soviet population and about 50 percent of Soviet industry after the Soviet Union had executed a surprise attack on the United

---

[346] Weigley, 444.

States.[347] This concept changed nuclear targeting strategy somewhat—it allowed striking of Soviet cities in a controlled discriminating way, while destroying as many military targets as possible.[348]

"Mutual assured destruction" (MAD) came into use later to describe the fact that both the U.S. and Soviet Union had the capability to inflict unacceptable damage on each other.[349] But this theory had a fatal flaw: MAD was based on McNamara's idea that the cost of nuclear war would be so great to both sides, that "deterrence simply would not fail."[350] His policy gave no credence to the possibility that it would fail, thus no specific guidelines were developed for different scenarios of employment of nuclear forces if deterrence failed.[351]

The weaknesses of McNamara's policy aside, the U.S needed the military capability to support this policy. Believing that bombers would not survive a Soviet first strike, McNamara allowed the strategic bomber force to decline in both numbers and capability.[352] He cancelled the new B-70 intercontinental bomber.[353] Instead, he built up land- and sea-based nuclear strike capability with Minuteman ICBMs and Polaris Submarine Launched Ballistic Missiles (SLBMs).[354] By 1968, the U.S. strategic missile force consisted of 54 Titan II and 1000 Minuteman land-based missiles and 656 Polaris SLBMs. Although the number of delivery vehicles did not increase until the 1980's, the U.S. started placing multiple warheads (multiple independently targetable reentry

---

[347] Paret, 757-758.
[348] Weigley, 444.
[349] _____, "Robert S. McNamara," *SecDef Histories*, 3.
[350] Paret, 758.
[351] Paret, 758.
[352] Stefan T. Possony, Jerry E. Pournelle, and Colonel Francis X. Kane, "Chapter 6, Assured Survival," *The Strategy of Technology* (electronic edition, 1997) 1; on-line, Internet, 15 September 2000, available from http://www.jerrypournelle.com/sot/sot_6.htm.
[353] Weigley, 447.

vehicles (MIRVs)) on missiles to enhance strike capability.[355] He also pulled overseas-based missiles.[356]

McNamara had such confidence in his belief that MAD would ensure deterrence never failed, he destroyed any hope of flexibility in employment of U.S. nuclear forces by limiting nuclear response options to massive nuclear strikes. Having ceded the initiative to the Soviets by declaring a no first strike policy, the U.S could only target cities and industrial areas, because any counterforce targets would already have been launched in a nuclear exchange. The Polaris missiles were not accurate enough for hitting anything but countervalue targets. Furthermore, none of the missiles in our inventory could be retargeted after launch. In essence, McNamara created the antithesis of flexible response options that he and Kennedy worked to create in 1961.[357]

But the worst part of McNamara's nuclear policy was his fundamental assumption that MAD would prevent a Soviet first strike. Deterrence based on MAD was only viable as long as the enemy believed it, too. But there is plenty of evidence that the Soviet Union never accepted McNamara's belief in MAD.[358]

The Soviets and "Mutual Assured Destruction"

Soviet parades of weaponry in Red Square during this time showed Soviet intent to build strong ABM defenses.[359] By 1963, the Soviets established their first ABM system: the 250 km-range SA-5 Griffon dual SAM/ABM system complex in Estonia, with 30

---

[354] _____, "Robert S. McNamara," *SecDef Histories*, 3.
[355] _____, "Robert S. McNamara," *SecDef Histories*, 3.
[356] Possony, Pournelle, and Kane, 3.
[357] Possony, Pournelle, and Kane, 3.
[358] Possony, Pournelle, and Kane, 3.
[359] William T. Lee, *Ballistic Missile Defense and Arms Control Follies,* 25 September 1996, 1-15; on-line, Internet, 9 January 2001, available from http://www.fas.org/spp/starwars/congress/1996_h/h960927l.htm.

other sites later added near Leningrad (current day St. Petersburg),[360] along with battle management radars.[361] The Soviets replaced it in the mid-1960's with the SA-5 Gammon.[362] The Soviets also built ABM system defenses around Moscow,[363] deploying the 322 km-range ABM-1 Galosh as a system of 64 launchers at 4 complexes by 1972.[364]

In addition, the Soviets had already started a decades-long program of constructing hardened shelters, command posts, and survivable communications networks for military and civilian leadership, coordinated wartime production, dispersal, and hardening plans for key industries, and developed evacuation procedures and shelters for people in urban areas.[365] Furthermore, the Soviets built hardened shelters for over 175,000 key government personnel. They also developed a redundant network of critical industrial and economic facilities, blast shelters for workers, and war mobilization plans for critical industries.[366]

More evidence comes from Soviet offensive nuclear missile deployment. Soviet nuclear strategy and force structure during this period were designed to destroy as many U.S. counterforce targets as possible, thereby limiting damage to the Soviet Union.[367] By 1966, the Soviets deployed the SS-9, a high accuracy (less than 1km circular error

---

[360] Vladimir Trendafilovski; *Russian Anti-Ballistic Guided Missile Systems: RZ-25 Anti-Ballistic Missile System,* 18 August 1998, on-line, Internet, 8 January 2001, available from http://www.wonderland.org.nz/rz-25.htm
[361] Lee, 1-8.
[362] Vladimir Trendafilovski; *Russian Anti-Ballistic Guided Missile Systems: SA-5 GRIFFON,* 18 August 1998, n.p.; on-line, Internet, 8 January 2001, available from http://www.wonderland.org.nz/rusabgm.htm.
[363] Lee, 1-8.
[364] Vladimir Trendafilovski; *A-35 Anti-Ballistic Missile System,* n.p.; on-line, Internet, 13 January 2001, available from http://www.wonderland.org.nz/a-35.htm.
[365] Defense Intelligence Agency, "Chapter III, Strategic Defense and Space Programs," in *Soviet Military Power 1985,* n.p.; on-line, Internet, 8 January 2001, available from http://www.fas.org/irp/dia/product/smp_85_ch3.htm
[366] Caspar Weinberger, Secretary of Defense, and George Schultz, *Soviet Strategic Defense Programs,* October 1985, n.p.; on-line, Internet, 13 January 2001, available from http://www.fas.org/irp/dia/product/ssdp.htm.
[367] Lee, 1-8.

probable (CEP), according to Western sources—Russian sources give a higher CEP of 1.3-1.9 km), high yield ICBM designed to destroy Minuteman Launch Control Centers (LCCs). But by 1969, the U.S. developed so much redundancy in Minuteman launch control (including an airborne system), that the Soviet Union deployed a MIRVed variant in the 1970's to counteract redundant LCCs.[368]

In the mid-1960's the Soviets also deployed the SS-11, an ICBM similar to the Minuteman. Its low accuracy (CEP of approximately 1.4 km) made it useful only for soft targets. In an effort to reach parity with the U.S., the Soviets deployed 990 of these missiles by 1972.[369]

Also during the 1960's, the Soviets were developing a system to overcome the West's geographic advantage of forward bases in Asia, Europe, and Turkey. Dubbed the Fractional Orbit Bombardment System (FOBS), it was "a modified upper stage launched by the SS-9 Mod 3, Scarp, carried a one- to three-megaton warhead and went into low-Earth orbit, giving the ICBM unlimited range and allowing it to approach the US from any direction, avoiding US northern-looking detection radars and, therefore, giving little or no warning."[370] This system, if used successfully, could have severely degraded our ability to retaliate: it could have destroyed our critical command and control networks,

---

[368] Federation of American Scientists, *R-36 / SS-9 SCARP,* 1; on-line, Internet, 10 January 2001, available from http://www.fas.org/nuke/guide/russia/icbm/r-36.htm.
[369] Federation of American Scientists, *UR-100 / SS-11 SEGO,* 1; on-line, Internet, 10 January 2001, available from http://www.fas.org/nuke/guide/russia/icbm/ur-100k.htm.
[370] Air University, "Chapter 1, Space History," in *A War Fighter's Guide to Space: Volume 1* (Maxwell AFB, AL: Air University Press, 1993), n.p.; on-line, Internet, 12 January 2001, available from http://www.fas.org/spp/military/docops/usaf/au-18/part01.htm.

and ABM radars, and strategic bombers before they could take off,.[371] (The Soviets phased out the orbital missiles by 1983 in compliance with the SALT II treaty.)[372]

McNamara and Anti-Ballistic Missile Capability

The Joint Chiefs of Staff, aware of the Soviets' growing capability in offensive and defensive nuclear weapons systems and passive defense measures, recommended procurement and deployment of the Nike-X anti-ballistic missile (ABM) system.[373] In 1965, the proposed Nike-X system consisted of 12 sites with 2 types of missiles (exoatmospheric and endoatmospheric) designed to intercept ICBMs and SLBMs.[374] The Nike-X combined an advanced radar system with a low-altitude intercept missile designed to intercept MIRVed reentry vehicles; furthermore, it could detect decoys and countermeasures. The only way to defeat the system was to send more "real" missiles against the system than it could intercept. This new system gave the U.S. a significant advantage over the Soviets, since their ABM systems could not defeat MIRVs.[375]

The research team tasked to analyze the Soviet threat and make recommendations briefed McNamara on their findings. First, the Soviets had a nuclear war fighting strategy diametrically opposed to McNamara's "mutual assured destruction" theory. The Soviets built their forces and strategy to minimize damage to their forces and population and to destroy our ability to retaliate—a counterforce strategy, not a countervalue strategy. Second, contrary to current National Intelligence Estimates (NIEs), the team concluded that the Soviets were developing MIRVed systems primarily for counterforce

---

[371] Federation of American Scientists, *R-36O / SL-X-? FOBS*, 1; on-line, Internet, 12 January 12 2001, available from http://www.fas.org/nuke/guide/russia/icbm/r-36o.htm
[372] Federation of American Scientists, *R-36O / SL-X-? FOBS*, 1.
[373] Lee, 3.
[374] Air University, "Chapter 1, Space History," *A War Fighter's Guide to Space: Volume 1*, n.p.
[375] Lee, 3.

targeting, and that deployment of the Nike-X and development of fallout shelters would reduce U.S. population fatalities far below McNamara's estimate of some 50 million casualties. Third, after careful study of the cost, the Army team concluded that deployment of a viable Nike-X ABM system would cost the U.S. about $4 for every $1 that the Soviets spent on offensive systems. This analysis was contrary to McNamara's own belief that a viable missile defense system would cost about $100 in defense for every $1 that the Soviets spent on offensive nuclear weapons. Therefore, the ABM system would be far more cost-effective than McNamara had anticipated.[376]

Although McNamara accepted the revised cost-effectiveness of an ABM system, he refused to accept the evidence of Soviet counterforce targeting strategy, saying "as a Soviet Marshall he would target the entire arsenal on U.S. cities."[377] This was more evidence of McNamara's unfortunate tendency to "mirror image" when considering enemy intentions and capabilities.

After the briefing, McNamara refused to authorize deployment of the Nike-X on the grounds that it was too "destabilizing," and would cause the Soviets to build up their inventory of MIRVed ICBMs. The Joint Chiefs took their case to President Johnson and pointed out several advantages of Nike-X deployment; namely, a) the Nike-X system would save millions of lives in a nuclear attack, b) if the Soviets did react by building up their MIRVed systems, they would have to divert massive funds from other military projects, and the nuclear yield on MIRVed warheads would be substantially reduced in any attack (e.g. 5 megatons on a MIRVed SS-18 vice 18-25 on a non-MIRVed RV), and

---

[376] Lee, 5-7.
[377] Lee, 7.

that the risk of Soviet attack would be reduced because the Soviets could not predict how well their attack would succeed against the Nike-X ABM system.[378]

Although McNamara originally agreed to deploy some limited ABM defenses, he quickly sabotaged the Nike-X deployment by attempting to discredit the system's effectiveness and minimizing the Nike-X deployment as to render it ineffective against a Soviet attack. By manipulating the variables in computer simulation of Nike-X effectiveness against a Soviet attack (no Soviet counterforce attacks, optimization of Soviet capability to exhaust all Nike-X interceptors) McNamara's systems analysts came up with the data McNamara wanted to support his "mutual assured destruction" position.[379]

Maintaining his position that an ABM system would be escalatory, McNamara suggested to Soviet Premier Alexei Kosygin at a joint U.S.-Soviet summit in June 1967 that both countries develop a treaty to enact strict limits on ABM system.[380] Kosygin responded that Soviet missile defensive systems around Moscow and Tallinn were designed to protect the population, and that before entering into any defense limitations, both sides should work on limiting offensive forces, first.[381] In essence, Kosygin admitted that the Soviets already had ABM systems deployed—in direct opposition to the CIA's assessment that the systems at Tallinn and Moscow were SAM systems. However, Kosygin's admission that the Soviets had an ABM system did not stop McNamara from his crusade to kill the Nike-X.

---

[378] Lee, 7-8.
[379] Lee, 8.
[380] Federation of American Scientists, *Anti-Ballistic Missile Treaty Chronology,* 1; on-line, Internet, 12 January 2001, available from http://www.fas.org/nuke/control/abmt/chron.htm
[381] Lee, 12.

With anti-military sentiment and military costs increasing due to the Vietnam War, McNamara was able to kill the Nike-X system in September 1967.[382] When McNamara did announce deployment of the Sentinel ABM system in 1967, it provided no protection from the Soviet threat. Rather, it was ostensibly designed to counter the much more limited threat from communist China.[383]

The Beginnings of SALT and the End of U.S. Strategic Superiority

By 1969, the Soviets had surpassed the U.S. in the number of land-based ICBMs deployed. The Nixon administration responded by authorizing replacement of the Sentinel ABM system with the Safeguard ABM system, and by MIRVing Minuteman 3 launchers. The Safeguard system (which was not actually deployed until the Ford administration) was designed to "protect land-based retaliatory forces from direct attack by the Soviet Union."[384]

In the face of these developments, the Soviets were ready to talk arms control. In November 1969, Strategic Arms Limitation Talks (SALT I) began. The resulting treaty, signed in 1972, limited both ABM and strategic nuclear offensive systems.[385] With this treaty, the U.S. ended any hope of maintaining strategic superiority over the Soviets, either in offensive or defensive strategic weapon systems.

The 1972 ABM Treaty initially limited strategic defenses to "200 launchers and interceptors, 100 at each of two widely separated deployment areas" to prevent "establishment of a nation-wide defense or the creation of a base for deploying such a

---

[382] Lee, 12,15.
[383] Lee, 12.
[384] Federation of American Scientists, *Anti-Ballistic Missile Treaty Chronology*, 1.
[385] Federation of American Scientists, *Anti-Ballistic Missile Treaty Chronology*, 1.

defense."[386] As amended by the 1974 protocol, it limited each side to one missile defense deployment area with no more than 100 launchers/missiles and guidance radars within a diameter of 150 kilometers, restricted early warning radars to the national periphery with only outward orientation, and prohibited nation-wide capability of non-nation-wide missile defense systems (such as theater missile defense systems) or transfer of missile defense components to foreign countries.[387] But as Soviet weapons deployments over the next 20 years later proved, this treaty was a ploy to deprive the U.S. of a means of defending itself from strategic attack, while giving the Soviets the time they needed to develop ABM capability.

SALT I was a boon for the Soviets, since the U.S. agreed to the number of Soviet offensive nuclear forces needed to make their nuclear war fighting strategy viable. In no way did it limit the Soviet's planned nuclear force arsenal—basically, SALT I gave them what they wanted.[388] By the end of the SALT I negotiations, the Soviets had 1500 land-based ICBMs, and their SLBM force had quadrupled. Furthermore, the huge payload on some Soviet ICBMs threatened our land-based missiles, even in hardened silos. The U.S. still had the same number of missiles as it had in 1967: 1054 ICBMs and 656 SLBMs.[389] Although the U.S. had more MIRVed warheads and more strategic bombers, with the Soviet's counterforce targeting strategy, their viability in a nuclear exchange was questionable.

---

[386] Federation of American Scientists, *Anti-Ballistic Missile Treaty Chronology,"*2.
[387] Council for a Livable World, *Nuclear Arms Control and the ABM Treaty,* 4; on-line, Internet, 11 January 2001, available from http://www.clw.org/coalition/nmdbook00abmtreaty.htm.
[388] Lee, 13.
[389] U.S. Department of State, *Strategic Arms Limitation Talks (SALT I)* n.p.; on-line, Internet, 12 January 20001, available from http://www.state.gov/www/global/arms/treaties/salt1.html

Overall, the strategic nuclear policies of the Johnson and Nixon administrations significantly damaged homeland defense capability. Their defense policies ended U.S. strategic superiority over the Soviet Union and made U.S. strategic forces vulnerable to a Soviet first strike. Furthermore, the ABM Treaty opened the door for the Soviet Union to develop a system of passive and active nuclear defenses that nullified McNamara's "mutual assured destruction" theory on which the ABM treaty was based.

The Effect of Vietnam on U.S. Homeland Defense

The damage to U.S. homeland defense capability was not limited to strategic nuclear offense and defense. McNamara's and Johnson's decision to escalate and micromanage a limited war of vague, unclear objectives to support a series of corrupt, unpopular, and incompetent South Vietnamese regimes against a communist insurgent force determined to win and backed by North Vietnam and China could arguably be considered the worst defense policy decision of the 20th century. As early as April 1965, then-Director of Central Intelligence John McCone warned:

> I think what we are doing in starting on a track which involves ground force operations...[will mean] an ever-increasing commitment of U.S. personnel without materially improving the chances of victory...In effect, we will find ourselves mired down in combat in the jungle in a military effort that we cannot win, and from which we will have extreme difficulty in extracting ourselves.[390]

Like his predecessor, Johnson did not understand the potential unintended consequences of his decision to escalate U.S. military involvement. In 1964, CIA analysts warned, "The costs of failure might be greater than the cost of failure under a counter-insurgency strategy because of the deeper U.S. commitment and the broader

---

[390] Center for the Study of Intelligence, *Episode 2, 1963-1965: CIA Judgments on President Johnson's Decision to "Go Big" in Vietnam,* n.p.; on-line, Internet, 12 February 2001, available from http://www.cia.gov/csi/books/vietnam/epis2.html.

world implications."[391] In other words, by escalating U.S. involvement, Johnson and McNamara put global U.S. credibility "on the line"; loss of this war would mean loss of our allies' confidence in our ability to contain communism and fight insurgencies, and would embolden our enemies to challenge U.S. influence, politically and militarily.

But President Johnson, while afraid of being blamed for losing South Vietnam, did not want to jeopardize funding for his "Great Society" domestic programs. So McNamara developed a war strategy that could be pursued cheaply while providing the illusion of winning: he developed a strategy of graduated escalation of military pressure on the enemy. Such a strategy would not win the war, but it would give the public the illusion of "not losing."[392]

Furthermore, McNamara relied on his civilian "whiz kids" to provide advice, instead of consulting with the Joint Chiefs. Hence, without forthright communication between the Joint Chiefs and McNamara, there was no possibility of reconciling McNamara's limited war effort and senior military officers' belief that the war could not be won with such a limited effort. U.S. conduct of the war degenerated into a total absence of real strategy—military activity (killing and counting enemy bodies, bombing targets in North Vietnam) substituted for strategy.[393] The Johnson administration's failure to successfully conclude the Vietnam conflict led to McNamara's resignation in 1967 and Johnson's decision not to seek reelection in 1968.[394]

---

[391] Center for the Study of Intelligence, *Episode 2, 1963-1965: CIA Judgments on President Johnson's Decision to "Go Big" in Vietnam*, n.p.
[392] H.R. McMaster, "Graduated Pressure: President Johnson and the Joint Chiefs," *Joint Forces Quarterly* (Autumn/Winter 1999-2000), 3; on-line, Internet, 12 February 2001, available from http://www.dtic.mil/doctrine/jel/jfq_pubs/1723.pdf.
[393] H.R. McMaster, 7.
[394] _____, "Robert S. McNamara," *SecDef Histories*, 6-7.

His successor, Richard Nixon, faced with an unpopular, expensive, and fruitless war, began a policy of Vietnamization: a gradual pullout of U.S. troops which forced the South Vietnamese to take on the responsibility of winning by themselves.[395] Vietnamization was an outgrowth of the Nixon Doctrine, a policy that required nations threatened by insurgencies and local wars to bear responsibility for their own defense. Under the Nixon Doctrine, although the U.S. would provide deterrent measures to prevent nuclear and conventional wars, U.S. assistance for governments facing insurgencies and local wars would be limited to material and economic assistance. The Nixon Doctrine was a direct result of America's entanglement in an unpopular war in Southeast Asia. Nixon was determined to prevent any further such involvements.[396]

In early 1973, Henry Kissinger signed a peace accord with the North Vietnamese, and the last American combat troops left South Vietnam in early 1973.[397] The peace agreement sealed South Vietnam's fate, for it allowed over 140,000 North Vietnamese troops to remain in South Vietnam. By 29 April 1975, the North Vietnamese captured Saigon.

The Vietnam War highlighted the critical importance of a clear national policy and strategy to execute the policy. In this case, defeat in a country of secondary importance eroded U.S. credibility as a champion of containment of communism and directly affected U.S. defense policy for decades to come.[398]

---

[395] Federation of American Scientists, *Vietnam War,* 1-8; on-line, Internet, available from http://www.fas.org/man/dod-101/ops/vietnam.htm.
[396] US Army Tank-automotive and Armaments Command (TACOM) Security Assistance Center, *The Nixon Doctrine,* 1-2; on-line Internet, 12 February 2001, available from http://www-acala1.ria.army.mil/tsac/nixon.htm.
[397] Federation of American Scientists, *Vietnam War,* 1-8.
[398] Federation of American Scientists, Vietnam War, 1-8.

Ford and Carter: Defense Nadir

President Ford inherited a nation angry and divided over the fallout from Vietnam and the Watergate scandal. America's credibility as a bulwark against communism was seriously damaged. Not only was U.S. conventional capability in question; U.S. capability to respond to a Soviet nuclear first strike was now questionable.

With the nuclear balance between the U.S. and U.S.S.R. now fundamentally changed for the worse since McNamara's tenure, Ford's Secretary of Defense James Schlesinger decided that U.S. strategic doctrine had to change with the new threat environment. He didn't agree with the doctrinal limitations of "assured destruction" and the implication that the only response to a Soviet nuclear strike would be wholesale destruction of Soviet cities. He was especially concerned that the Soviets had the capability of attacking U.S. and European cities even after a U.S. retaliatory strike. Unlike McNamara, Schlesinger believed that there was a chance that deterrence would fail, and he wanted more than one option to respond. Schlesinger wanted flexibility to respond selectively to any attack to limit further escalation and prevent collateral damage as much as possible. Schlesinger thus introduced the concept of flexible strategic targeting.[399]

Schlesinger Tries to Undo the Damage

To carry out his concept, he called for more research and funding to produce maneuverable warheads and new ICBMs for counterforce strikes. In response to growing Soviet counterforce capabilities, Schlesinger urged that the U.S. maintain "essential equivalence in terms of strategic nuclear forces so that "everyone…will perceive that we

---

[399] Douglas Kinnard, *The Secretary of Defense* (Lexington, Kentucky: The University Press of Kentucky, 1980), 174-177.

are the equal of our strongest competitors."[400] In keeping with this goal, he pushed Congress to maintain funding for the new B-1 strategic bomber (the proposed replacement for the aging B-52s). The B-1 would force the Soviets to reallocate their defense resources against this new strategic asset; the B-1 was also designed to counteract the Soviets' numerical advantage in strategic missiles.[401]

Schlesinger was also concerned over the deficiencies in U.S. conventional force capability. He argued that any previous advantage the U.S. had over the Soviet Union in strategic forces had vanished, thereby erasing any inhibition the Soviets previously had in exploiting conventional capability. He argued that by maintaining weakened conventional forces, the U.S. and NATO lowered the threshold for nuclear exchange.[402] But in the end Schlesinger lost the battle for improvements in both conventional and strategic forces; with Vietnam still a fresh memory and continuing economic problems, Congress would not fund expensive defense programs.

Although Scheslinger managed to keep the B-1 program afloat during his brief tenure, he was not able to convince either Congress or President Ford of the need to increase defense spending to provide a viable deterrent conventional or strategic force against growing Soviet capabilities and to maintain U.S. preeminence as a guarantor of defense of the free world. He was outraged over Congress's proposed $5 billion cut in the fiscal year 1976 defense budget, which would require a 200,000-person cut in active duty military strength. With the U.S. in the middle of severe economic problems and a Presidential election looming the next year, Ford's advisors convinced him to run on a

---

[400] Kinnard, 177-178.
[401] Kinnard, 183-184.
[402] Kinnard, 178-182.

balanced budget platform, which meant accepting deep cuts in defense. Schlesinger's refusal to support the proposed defense cuts resulted in his forced resignation.[403]

Effects of Recession on Homeland Defense

Schlesinger was right in his policies, as later events bore out. But in the post-Vietnam era, severe economic problems caught center stage, and neither Ford nor Carter supported large increases in defense spending. By 1980, inflation rose to 16 percent, interest rates reached 20 percent, and the budget deficit tripled to $75 billion.[404] Faced with anti-military and anti-interventionist public sentiment, inflation, and recession, defense spending sharply declined throughout the 1970's.[405] As a percentage of the Gross National Product (GNP), the defense budget fell from 9.4 percent in 1968[406] to 5.3 percent in 1976 (and in 1981).[407]

The fallout from lack of Congressional and Presidential support for defense improvements in the 1970's had a terrible effect on U.S. strategic and conventional defense capability. The U.S. ended up with no ABM system, the B-1 bomber was cancelled, and the military became a "hollow force."

ABM capability was one of the first programs to fall. In accordance with the provisions of the ABM treaty, in 1975 President Ford had authorized deployment of a single Safeguard ABM system with 100 launchers, interceptors, and associated radars at

---

[403] Kinnard, 183-187.
[404] Dr. Steve Schoenherr, University of San Diego History Department, *Ford-Carter Era: 1974-1980,* 3; on-line, Internet, 15 January 2001, available from http://history.acusd.edu/gen/20th/carter.html.
[405] Jordan, Taylor, and Korb, 73.
[406] Jordan, Taylor, and Korb, 73.
[407] Dennis S. Ippolito, "Defense Budgets and Spending Control: The Reagan Era and Beyond," in *Defense Policy in the Reagan Administration,* ed. William P. Snyder and James Brown (Washington, D.C.: National Defense University Press 1988), 172.

Grand Forks, North Dakota. But the high cost of maintaining the system, plus its questionable capability, led to its deactivation in 1976.[408]

Carter and the B-1

During this period, the U.S. did not improve its strategic offensive forces, either. Carter cancelled the new B-1 bomber in 1977, depriving the U.S. of a long-range strategic bomber capable of penetrating Soviet air defenses, leaving the U.S. with its aging fleet of B-52s.[409] Instead, he decided to equip B-52s with air-launched cruise missiles, to be launched before the bombers reached Soviet air space. The problem with this strategy was that, assuming the bombers could survive a Soviet first strike, Soviet interceptors would certainly attack the B-52s before they could reach projected launch areas.[410]

Effects of Conventional Force Cuts

Carter also cut Ford's five-year program to build 157 Navy ships by 50 percent, and he cancelled a new nuclear-powered aircraft carrier in 1979.[411] Sea-based airpower continued to decline, and by 1977, the U.S. had only 13 aircraft carriers to patrol the world's oceans. Loss of carriers caused a drastic change in naval defense strategy. Since carrier maintenance and rotation requirements demand three carriers to maintain a one-carrier presence in a specified area of operations, with only 13 carriers, only 4 were available from any time overseas. Two were based in the Pacific (with the added responsibility for the Indian Ocean) and two were stationed in the Mediterranean, with

---

[408] Federation of American Scientists, *Anti-Ballistic Missile Treaty Chronology*, 2.
[409] Federation of American Scientists, *B-1A,* 25 March 1998, 1; on-line Internet, 26 January 2001, available from http://www.fas.org/nuke/guide/usa/bomber/b-1a.htm.
[410] John M Collins, *American and Soviet Military Trends Since the Cuban Missile Crisis* (Washington, D.C.: The Center for Strategic and International Studies, Georgetown University 1978), 109.
[411] Korb, 61.

one of these "on call" for duty in the North Atlantic.[412] The Navy had to change its strategy to "defensive sea control"; instead of forward-based deterrence, the Navy had to give priority to "barrier operations and…close-in defense of the sea lanes."[413]

Soviet Military Buildup

Conversely, the Soviets increased defense spending by about 5 percent per year throughout the 1970's. During this decade, the Soviets spent $104 billion more on defense than the U.S. on new weapons systems, producing "six times as many tanks, twice as many combat aircraft, and three times as many ships, while developing twice as many new strategic systems."[414] Between 1970 and 1977, the Soviets built 47 new nuclear submarines,[415] added 589 SLBMS to their inventory,[416] and surpassed the U.S. in total ICBM warheads by 1977.[417]

The overall effect of this defense imbalance was strategically devastating for the U.S. American ICBMs were at risk in a first strike, the U.S. Navy no longer maintained unchallenged control of the sea, and both the U.S. and NATO countries now were vulnerable to attack by the Soviet Union and Warsaw Pact countries.[418]

The Iranian Hostage Crisis

Our enemies also knew we were vulnerable, and took advantage of the situation. In November 1979, Iranian "students" took over the U.S. Embassy in Tehran and seized 66

---

[412] Collins, 251.
[413] John Allen Williams, "The US Navy Under the Reagan Administration and Global Forward Strategy," in *Defense Policy in the Reagan Administration*, ed. William P. Snyder and James Brown (Washington, D.C.: National Defense University Press 1988), 277.
[414] Lawrence J. Korb, "The Defense Policy of the United States," in *The Defense Policies of Nations: A Comparative Study*, ed. Douglas J. Murray and Paul R. Viotti (Baltimore, MD: Johns Hopkins University Press 1982), 52.
[415] Collins, 102-103.
[416] Collins, 98.
[417] Collins 94.
[418] Korb, 52-53.

hostages, beginning a hostage crisis that did not end until the day President Reagan was inaugurated.[419] The Sandinistas defeated Somoza in Nicaragua in August 1979, sparking fears of another Cuba in the Western hemisphere.[420] The Russians then invaded Afghanistan in December 1979, ending any chance of U.S. Senate ratification of the SALT II arms limitation treaty with the Soviet Union.[421]

The End of SALT II

Carter was furious over what he considered to be a Soviet betrayal, especially in light of the recently concluded SALT II nuclear weapons limitation talks. His fury betrayed his naiveté regarding the Soviets' ultimate intentions when he said, "This action of the Soviets has made a more dramatic change in my own opinion of what the Soviets' ultimate goals are than anything they've done in the previous time I've been in office."[422]

The Carter Doctrine

In response to the Soviet invasion, and concerned about a potential Soviet incursion into the oil-rich Middle East, Carter warned in his 1980 State of the Union address, "…an attempt by any outside force to gain control of the Persian Gulf region will be regarded as an assault on the vital interests of the United States of America. And such an assault will be repelled by any means necessary, including military force."[423] In what became known as the "Carter Doctrine," President Carter completely reversed the direction of the Nixon

---

[419] Schoenherr, 3.
[420] Peter R. Zwick, "American-Soviet Relations: The Rhetoric and Realism," in *Defense Policy in the Reagan Administration*, ed. William P. Snyder and James Brown (Washington, D.C.: National Defense University Press 1988), 88.
[421] Jordan, Taylor, and Korb, 334.
[422] John Newhouse, *War and Peace in the Nuclear Age* (New York: Alfred A. Knopf, 1988), 331.
[423] U.S. Army Tank-automotive and Armaments Command (TACOM) Security Assistance Center, *The Carter Administration,* 2; on-line, Internet, 12 February 2001, available from http://www-acala1.ria.army.mil/tsac/carter.htm.

Doctrine regarding security assistance. This was the first statement by a President since Vietnam of possible use of U.S. troops to protect vital interests—in this case, Persian Gulf oil supplies.[424]

But with the relative weakness of the U.S. military at the time, his ability to execute this doctrine, if required, was questionable. With the Soviet military in Afghanistan, the Soviets had the upper hand in a potential Middle East conflict. They could use their bases in Afghanistan to stage Backfire bomber attacks on U.S. naval forces in the Arabian Sea; furthermore, the bases in Afghanistan enhanced the range of fighter protection for Backfire bombers in a potential conflict.[425]

Desert One: Nadir of U.S. Defense Credibility

Carter scrambled to restore U.S. strategic deterrent credibility in the Middle East. On 1 March 1980, he established the Rapid Deployment Joint Task Force (RDJTF), to be used for missions to the Middle East and Southwest Asia. Less than two months later, he attempted to rescue the American hostages in Iran.[426] In the ensuing disaster, America's loss of strategic superiority was evident for the entire world to see.

The failed hostage rescue attempt in Iran that ended in flames at Desert One in April 1980 best symbolized America's loss of defense capability.[427] Many factors contributed to the failure. The wrong aircraft were used for the mission; Sea Stallions are used for

---

[424] U.S. Army Tank-automotive and Armaments Command (TACOM) Security Assistance Center, *The Carter Administration*, 2.
[425] Raymond Tanter, "Chapter Two: Iran: Balance of Power vs. Dual Containment," in *Rogue Regimes: Terrorism and Proliferation* (New York: St. Martins Press, September 1996) 13; on-line, Internet, 12 February 2001, available from http://www-personal.umich.edu/~rtanter/rogue.iran.html.
[426] Lawrence E. Grinter, *Avoiding the Burden: The Carter Doctrine in Perspective,* 2; on-line, Internet, 12 February 2001, available from www.airpower.maxwell.af.mil/airchronicles/aureview/1983/jan-feb/grinter.html.
[427] John F. Guilmartin, Jr., "Terrorism: Political Challenge and Military Response," in *Defense Policy in the Reagan Administration*, ed. William P. Snyder and James Brown (Washington, D.C.: National Defense University Press 1988), 116.

mine sweeping at sea level and are not designed for flying over mountainous terrain in sandstorms. Only six of the original eight helicopters made it to Desert One (the refueling point), and one malfunctioned at Desert One, leaving the rescue force no choice but to declare a no-go for the mission, since there weren't enough helicopters left to carry the troops, hostages, and fuel needed to escape Tehran. Furthermore, the security imposed by the Carter administration was such that the entire rescue force had never had a full rehearsal of the mission prior to execution, and no outside group was allowed to review the plan prior to execution—basically, the planners reviewed their own plans. The Joint Chiefs of Staff approved the plan in the mistaken belief that a full dress rehearsal had been conducted.[428] These factors culminated in the crash of a helicopter into one of the transport aircraft at Desert One—eight soldiers died and America was humiliated. The failure at Desert One was probably the single greatest contributing factor to Carter's defeat in the 1980 election.

The Ford-Carter Era: Lessons Learned

The 1970's ended with American defense credibility at its nadir. We had lost strategic superiority in the Middle East. Elsewhere in the Middle East, Africa, and Latin America, Marxist governments had taken over strategically important countries (including Angola, Mozambique, Ethiopia, South Yemen, Cambodia, Laos, Rhodesia, Afghanistan, Nicaragua)[429] and were exporting leftist revolution to their neighbors. America's loss in Vietnam, loss of nuclear superiority over the Soviet Union, the collapse of arms control negotiations, and finally, the Desert One disaster clearly showed erosion of U.S. homeland defense capability. The Soviet Union now had a credible chance of a

---

[428] David C. Martin and John Walcott, *Best Laid Plans: The Inside Story of American's War Against Terrorism* (New York: Harper and Row, 1988), 11-35.

successful nuclear first strike; our strategically important sea lines of communication for oil and other critical resources in both the Middle East and the Caribbean were no longer secure from hostile nations backed by the Soviet Union and Cuba; governments in Africa, southeast Asia, and Latin America were losing ground to leftist insurgencies, thereby threatening U.S. access and influence to important resources and lines of communication. This situation propelled Ronald Reagan into the White House with his promise to restore America's defense capability.

Reagan and Bush: Defense Resurgence

Defense Buildup and the End of Detente

When President Reagan entered the White House in 1981, he drastically changed U.S. defense policy. Reagan, determined to prevail over the Soviet Union, abandoned détente in favor of a defense buildup. Reagan felt détente had failed to enhance U.S. security or mitigate tensions between the U.S. and Soviet Union because détente ignored Soviet expansionism and failed to redress the Soviets' upper hand in nuclear weapons.[430] Hence, Reagan initiated the largest peacetime military expansion in U.S. history as the basis for his defense policy. His goal was to fundamentally alter the strategic and conventional balance between the U.S. and Soviet Union so that the U.S. could then confront the Soviet Union from a position of strength.[431]

Reagan's National Security Strategy

Reagan and his staff first developed a national security strategy that went beyond containment—instead, he was determined to reverse Soviet strategic gains. To achieve

---

[429] Grinter, 1-2.
[430] William P. Snyder and James Brown, ed., "Introduction," in *Defense Policy in the Reagan Administration*, (Washington, D.C.: National Defense University Press 1988), xv.
[431] Jordan, Taylor, and Korb, 78.

his goal, his National Security Strategy of 1982 emphasized a key new objective: "*contain and reverse the expansion of Soviet control and military presence throughout the world* (italics added), and to increase the costs of Soviet support and use of proxy, terrorist, and subversive forces."[432] Reagan's idea for a national security strategy was simple and unequivocal: "We win, they lose."[433]

Conventional Forces Buildup

President Reagan was determined to develop a force structure that could support his new National Security Strategy.[434] Conventional forces doctrine changed to a more aggressive, offensive stance, and military force structure changed to accommodate new doctrines.

To reclaim maritime superiority, the Reagan administration increased the Navy's fleet from 479 (number of warships as of 1980) to 567 warships (including two new aircraft carriers) by 1987[435] and changed naval maritime strategy back to an offensive emphasis.[436] This new strategy, called the "Maritime Strategy," emphasized offensive strikes against Soviet forces, even in their own littoral. This meant that the Navy had to have a force structure that enabled our forces to operate north of the Greenland-Iceland-Norway line to prevent Soviet fleet assets from operating further south in event of a war with NATO.[437] This new strategy was diametrically opposed to the Carter administration

---

[432] President, *National Security Decision Directive Number 32, U.S. National Security Strategy* (20 May 1982): 1; on-line, Internet, 17 January 2001, available from http://www.fas.org/irp/offdocs/nsdd/23-1618t.gif.
[433] Quoted by Jonathan Fox, *United States Foreign Policy in the Twenty-First Century: The Crisis and Renewal of the Republican Empire*, 5; on-line, Internet, 9 February 2001, available from http://www.spaef.com/JPE_PUB/vln3_fox.html.
[434] Snyder and Brown, "Introduction," xv.
[435] John Allen Williams, "The U.S. Navy Under the Reagan Administration," in *Defense Policy in the Reagan Administration*, (Washington, D.C.: National Defense University Press 1988), 291.
[436] John Allen Williams, 289.
[437] John Allen Williams, 278-279.

strategy of "defensive sea control," which Secretary of the Navy John Lehman justly criticized as a "Maginot Line" defeatist strategy.[438]

The Army also developed a new, aggressive doctrine—AirLand Battle Doctrine, designed to defeat the Soviet Army in a large-scale conventional war. The new doctrine relied on rapid offensive action—seizing and maintaining the initiative, maneuvering and striking before the enemy could react while denying the enemy reinforcement.[439] In addition to conventional war capability, the Army also improved its capacity for limited war and counter-terrorism operations. To meet requirements for low intensity "come as you are" conflicts, the Army increased its light infantry divisions to enable the Army to quickly deploy to problem areas before conflicts expanded into larger, conventional wars. This was accomplished without increasing active-component end-strength, enabling the Army to pursue its badly needed equipment modernization programs. Several important missions were transferred to the reserve component—by 1987, reserve components provided 50 percent of the Army's Special Forces and 90 percent of its psychological operations and civil affairs units.[440]

In an effort to meet the growing threat of terrorism, the Army also enhanced its Special Operations Forces (SOF) as part of an overall special operations capability buildup across the DOD. Across the services, the Reagan administration increased SOF from 10,000 to 15,000 and added $200 million to the special operations forces budget.[441]

---

[438] John Allen Williams, 278.
[439] William O. Staudenmaier, "The Decisive Role of Landpower," in *Defense Policy in the Reagan Administration*, (Washington, D.C.: National Defense University Press 1988), 260.
[440] Staudenmaier, 260-262.
[441] Guilmartin, 133-134.

Although the Air Force's strategic attack capability was enhanced by reinstating the B-1 bomber program,[442] unfortunately, the Air Force did not follow suit in terms of prioritization of counter-terrorism capability—despite the fact that terrorism had become one of the greatest threats to U.S. civilians and forces world-wide. By 1986, two Senators glumly noted that the Air Force had the same number of MC-130 and AC-130 A/C gunships, and two less HH-53 Pave Low helicopters than existed at the time of Desert One in 1980.[443] Air Force leadership argued that Air Force systems had to be capable of fighting across the spectrum of conflict. Given this attitude, specialized aircraft for inserting and supporting SOF on clandestine or covert missions were bound to get short shrift in funding and senior support.[444]

International Terrorism and Homeland Defense

The 1979 Iranian hostage crisis, 1981 kidnapping of Brigadier General James Dozier, 1983 bombing of the Marine barracks in Beirut and the U.S. Embassy in Beirut, 1984 bombing of the U.S. Embassy Annex in Beirut, 1985 Rome and Vienna airport massacres, 1985 hijacking of TWA Flight 847 and subsequent murder of Navy diver Robert Stetham, April 1986 bombing of the La Belle discothèque in Berlin, and murders of several American military and diplomatic representatives throughout the 1980's[445] all underscored the fact that America faced a new kind of warfare—international terrorism. Clearly, America needed a new strategy and force structure to counter this new threat that defied defeat by conventional force capability.

---

[442] Schuyler Foerster, "Arms Control: Redefining the Agenda," in *Defense Policy in the Reagan Administration*, (Washington, D.C.: National Defense University Press 1988), 24-25.
[443] Thomas A. Fabyanic, "The U.S. Air Force," in *Defense Policy in the Reagan Administration*, (Washington, D.C.: National Defense University Press 1988), 326.
[444] Fabyanic, 325.
[445] Martin and Walcott, xv-xxii.

Congress became so concerned about funding and force structure for counter-terrorism capability, that in October 1986, Congress created a new unified command—United States Special Operations Command (USSOCOM). Congress granted USSOCOM authority to develop and acquire equipment unique to SOF needs to ensure that its force structure would not fall victim to other priorities.[446] With this action, Congress supported SOF structure, training, and equipment needs to meet the newest threat to America—international terrorism.

But despite improvements in counter-terrorism force structure, execution of public counter-terrorism policy did not fare as well. Despite Reagan's announced policy of "swift and effective retribution"[447] for terrorist acts against Americans, in actual execution, the Reagan administration was inconsistent. One of the problems he faced was lack of the necessary intelligence apparatus to provide timely, correct intelligence about terrorists. Nothing had been done to improve U.S. human resources intelligence (HUMINT) capability in the 1970's, and lack of resources handicapped counter-terrorism efforts.[448]

Furthermore, the Reagan administration's actions in response to some terrorist incidents were confusing at best, in light of his stated policy. During the TWA Flight 847 hostage crisis, the administration was stymied by disagreements between the NSC and the rest of the government over the appropriate course of action. He also made the same mistake as the Carter administration had during the Iran crisis by essentially ruling out use of force in a public statement—thereby allowing the hijackers to control the course of events. Eventually the TWA hostages were released—when Israel agreed to release

---

[446] Fabyanic, 326-327.
[447] Martin and Walcott, 43.

Shiite prisoners in exchange for the hostages and after Syrian President Hafez Assad brokered the deal with Iran and Hezbollah. In essence, the Reagan administration broke its own pledge of "no deals" with terrorists to win the release of the hostages.[449]

In yet another concession to terrorism, Reagan's staff brokered an arms-for-hostages deal with Iran from 1985 to 1986 to obtain release of the five American hostages held by Hezbollah in Lebanon. Middle Eastern terrorist groups were quickly learning that there was something to be gained by holding Americans hostage.[450]

Soon after the release of the five American hostages from Lebanon, Palestinian terrorists hijacked the *Achille Lauro*, demanding release of 50 Palestinians held in Israeli prisons. Although the Egyptian government and Yassar Arafat managed to obtain the release of the hostages without release of the Palestinian prisoners, the combined efforts of the Egyptian and Italian governments ensured that the mastermind of the hijacking, Abu Abbas, escaped. However, the masterful intercept by U.S. Navy F-14s of the aircraft carrying the *Achille Lauro* hijackers ensured the four hijackers went to prison.[451] The qualified success of the *Achille Lauro* outcome gave the U.S. government new confidence in use of military force against international terrorism.

The greatest test of U.S. counter-terrorism capability came soon afterward. In November 1985, the Abu Nidal terrorist group, sponsored by Libya's Muammar Qaddafi, hijacked an EgyptAir Flight as it took off from Athens to Cairo. Sixty people died from the explosion and fire when Egyptian commandos stormed the plane in a horribly botched rescue attempt. Next, the Abu Nidal group, again backed by Qaddafi, executed

---

[448] Martin and Walcott, 47-48.
[449] Martin and Walcott, 195-202.
[450] Martin and Walcott, 229-234.
[451] Martin and Walcott, 235-257.

two massacres in December 1985 at the Rome and Vienna airports. Following these attacks, in response to confrontations with the U.S. Navy in the Gulf of Sidra, Qaddafi ordered agents in his Libyan People's Bureaus to start attacking U.S. military installations and civilian targets frequented by Americans. On 5 April 1986, Libyan agents detonated a bomb in Berlin's La Belle discothèque, a nightclub frequented by Americans. Two American soldiers and a Turkish woman were killed.[452]

Reagan and his staff decided to retaliate with a conventional attack to try to destroy Qaddafi's capability to support terrorism. Although Iran and Syria were responsible for more attacks and fatalities of Americans, Reagan chose to go after Libya. Qaddafi was more vulnerable to military attack than Iran or Syria: unlike Syria, Libya had no friendship treaty with the Soviets to protect it; unlike Iran, the U.S. wasn't making a secret deal with the government to obtain release of hostages. Also, Qaddafi's terrorism apparatus was an immediate and real danger to Americans.[453]

Britain was the only European country to support the U.S. raid. Prime Minister Margaret Thatcher, who owed Reagan for his support of Britain during the Falklands War, allowed U.S. F-111s to take off from Britain. But the F-111s had to fly around the European continent since no other country would give the U.S. overflight rights for the raid. The distance to target was far beyond the normal limits of an F-111; even on a two and one-half hour sortie, 40 percent of the aircraft could be expected to malfunction. The restrictive rules of engagement for the raid, designed to minimize civilian casualties, plus the effects of the flying distance on the F-111s (some of the aircraft never made it to their targets), minimized actual damage to the targets. The U.S. raid inflicted some damage on

---
[452] Martin and Walcott, 258-288.
[453] Martin and Walcott, 285-315.

Libyan terrorist infrastructure, but did not destroy Libya's ability to sponsor terrorist acts. Several Americans and British, including three hostages in Lebanon, were killed in retaliation.[454] Libya's December 1988 bombing of Pan Am Flight 103 was quite probably in retaliation for the raid.

The Reagan administration showed only too clearly the difficulty in developing and executing effective policy for terrorism. Concessions encouraged more terrorist acts, and reprisals invited reprisals. Reagan's counter-terrorism policy did not achieve any lasting improvement in defense of U.S. vital interests against terrorists and their state sponsors because it was inconsistent and sporadic, and did not maintain a consistent, sustained course of action using all instruments of national power to isolate and destroy both terrorist groups and their state sponsors.

Security Assistance and Homeland Defense

Another area homeland defense policy that produced only mixed results was Reagan's security assistance policy. Regan provided various forms of economic and military aid to anti-communist groups throughout his Presidency. In his February 1985 State of the Union address, President Reagan pledged:

> We must not break faith with those who are risking their lives…on every continent from Afghanistan to Nicaragua…to defy Soviet aggression and secure rights that have been ours from birth. *Support for freedom fighters is self-defense* [italics added].[455]

The Reagan Doctrine

This policy became known as the Reagan Doctrine: the use of various forms of economic and security assistance to reverse Sovietization of countries in Asia, Africa,

---

[454] Martin and Walcott, 285-315.

and Latin America. In contrast to the Carter Doctrine, which emphasized prevention of Soviet control over the Middle East, Reagan's policy sought to reverse Soviet gains in Marxist countries by actively supporting anti-communist insurgent movements.[456] Reagan initiated these security assistance programs in an attempt to change the strategic balance of power by eroding Soviet influence in strategically critical countries.[457]

The Reagan Doctrine was not an unqualified success by any standard. In Nicaragua, the doctrine was used to train and arm a motley collection of anti-Sandinista forces, some of which were not necessarily proponents of freedom and democracy. By 1985, the United States had spent over $100 million dollars to undermine the Sandinista government.[458]

But the policy turned out to be an embarrassment for the Reagan administration for several reasons. First, Congress vacillated throughout the 1980's on the level and type of support for Nicaraguan rebels and openly disagreed with Reagan and his policy advisors on the type and amount of aid to give to the rebels. Second, the goals of Reagan's policy were not clear (overthrow the Sandinista regime, or just try to convince Ortega to stop exporting leftist revolution to neighboring countries and allow more freedom within Nicaragua?). And third, the Iran-Contra scandal (profits from selling arms to Iran were illegally diverted to assist Nicaraguan rebels) displayed our disjointed and uncoordinated U.S. policy on Nicaragua and initiated a Congressional investigation, causing

---

[455] Ronald Reagan, quoted in *Reagan and the Soviets: Reagan Doctrine,* 1; on-line Internet, 12 February 2001, available from http://www.reagan.dk/newreadoc.htm.

[456] _____, *Reagan and the Soviets: Reagan Doctrine,* 1; on-line Internet, 12 February 2001, available from http://www.reagan.dk/newreadoc.htm.

[457] Michael Mandelbaum, "American Policy: The Luck of the President," *Foreign Affairs* 64, no. 3 (1986): 411.

[458] Ted Galen Carpenter, "U.S. Aid to Anti-Communist Rebels: The "Reagan Doctrine" and Its Pitfalls," *Policy Analysis* No. 74 (24 June 1986): 15, 19; on-line, Internet, 12 February 2001, available from http://www.cato.org/pubs/pas/pa074.html.

international embarrassment for Reagan. In the end, a peace plan brokered by Costa Rican President Arias and supported by neighboring Central American states brought an end to the fighting in 1988, and free elections were held in Nicaragua in 1990.[459]

The Reagan Doctrine as applied in Nicaragua was not a success in supporting homeland defense. Although U.S. support for the rebels undoubtedly contributed to Daniel Ortega's decision to agree to a cease-fire in 1988, the policy did not result in toppling the Sandinista regime, the years of policy vacillations portrayed America as unreliable and inconsistent in its will to support the anti-Marxist insurgencies, and illegal actions in support of the rebels caused a major embarrassment to the President and his advisors.

The Reagan Doctrine applied in Afghanistan is also a cautionary tale concerning the unintended effects of security assistance on homeland defense. Although U.S. assistance helped drive the Soviets out of the country, in the long-term it created far greater problems.

Afghanistan has long been a country divided by ethnic, tribal, and religious differences. As a result, Afghanistan has a long history of political instability and revolts. After a 1978 coup that brought Hafizullah Amin to power in Afghanistan, the Soviets constantly provided assistance to Amin's government to fight a growing insurgency of Afghan resistance groups. But Amin refused to take Soviet "advice" on handling the increasingly unstable internal security situation. To protect their power base in this strategically important country, the Soviet Union invaded Afghanistan in December 1979 and installed Babrak Karmal as Prime Minister of a communist puppet government. But

---

[459] James M. Scott, "Interbranch Rivalry and the Reagan Doctrine in Nicaragua," *Political Science Quarterly*, Summer 1997, 1-10; on-line, Internet, 16 February 2001, available from

the vast majority of Afghans opposed Karmal's government, and the seven principal Afghan tribal guerrilla groups (collectively called the "mujaheddin") that opposed Karmal formed an alliance of convenience to coordinate their war against the Soviet occupation forces and oust Karmal.[460]

Concerned that the Soviets could use Afghanistan to stage attacks in the Middle East and Indian Ocean, Reagan continued his predecessor's policy of providing covert military aid via Pakistan in support of the mujaheddin.[461] Although arms and other assistance from the U.S., Pakistan, and other sympathetic countries helped turn the war around in Afghanistan, this assistance helped create a significant, long-term threat to the U.S. homeland and its allies.

The 1989 Soviet withdrawal from Afghanistan created a power vacuum in its aftermath. The agreement between Afghanistan, Pakistan, the Soviet Union, and the U.S. that led to the 1989 departure of Soviet troops did not include the mujaheddin either in the negotiations or the agreement. As a result, they refused to accept the accords. So, instead of ending the civil war with the withdrawal of Soviet troops, the agreement helped escalate the civil war. The titular head of Afghanistan's government, Muhammad Najibullah (the brutal former chief of the Afghan secret police), had been put in place by the Soviets to replace Karmal in 1986. He had no popular support, and after the Soviets left, his government collapsed in 1992.[462]

The civil war escalated between ethnic divisions within both the Afghan army and the mujaheddin. Even as the mujaheddin entered Kabul in 1992, a new round of ethnic

---

www.britannica.com/bcom/magazine/article/0,5744,237514,00.html.
[460] Federation of American Scientists, *Afghanistan—Introduction*, 2-7; on-line, Internet, 17 February 2001, available from http://www.fas.org/irp/world/afghan/intro.htm.
[461] Carpenter, 4, 11.

fighting erupted between various mujaheddin groups, each bent on control of Afghanistan. By 1994, Kabul was divided, with different parts of the city under the control of different ethnic and religious factions. What was left of the Afghan central government collapsed, and the country was in anarchy.[463]

From the chaos, one ethnic group emerged as the most powerful—the ultra-conservative Islamic movement known as the Taleban. By the end of 1997, it controlled two-thirds of the country and Kabul, with other factions fighting to maintain control of what was left in the northern provinces.[464] The Taleban established a de facto government, repressive even by Southwest Asia standards, and became the world's second-largest producer of opium—now the mainstay of what is left of Afghanistan's economy.[465]

The Taleban also became host to Islamic extremists from around the world, allowing them to use Afghanistan for training and basing of worldwide terrorist operations.[466] It also now provides safe haven to the man who is arguably the single greatest threat to the U.S. and its interests: Usama bin Laden. Usama bin Laden developed an organization in the 1970's to recruit Muslim fighters in their war against the Soviets in Afghanistan. In the late 1980's he formed a terrorist network, and after his expulsion from Saudi Arabia in 1989 and Sudan in 1996, he came to reside in Afghanistan.[467]

---

[462] Federation of American Scientists, *Afghanistan—Introduction*, 6-7.
[463] Federation of American Scientists, *Afghanistan—Introduction*, 6-7.
[464] Federation of American Scientists, *Afghanistan—Introduction*, 7.
[465] Federation of American Scientists, *Afghanistan—Introduction*, 1.
[466] U.S. Department of State, Coordinator for Counter-terrorism, *Patterns of Global Terrorism 1999*, (Washington, D.C.: U.S. Government Printing Office, April 2000), 7-8; on-line, Internet, 16 February 2001, available from http://www.state.gov/www/global/terrorism/1999report/asia.html#Afghanistan.
[467] U.S. Department of State, Office of the Coordinator for Counter-terrorism, *Fact Sheet: Usama bin Laden*, 21 August 1998,1-2; on-line, Internet, 16 February 2001, available from http://www.state.gov/www/regions/africa/fs_bin_ladin.html.

From Afghanistan, Usama bin Laden now directs and funds several Islamic extremist groups that commit terrorist attacks worldwide. His network supports terrorists in Egypt, Afghanistan, Bosnia, Chechnya, Tajikistan, Somalia, Yemen, and Kosovo, and trains terrorists in the Philippines, Algeria, and Eritrea. In August 1996, bin Laden issued a "declaration of war" against the U.S., and in 1998 declared his intention to kill Americans and our allies anywhere in the world.[468] He was responsible for the 1998 U.S. Embassy bombings in Kenya and Tanzania[469] and was probably the mastermind of the bombing of the U.S.S. Cole.[470]

The situation in Afghanistan is another example of a security assistance program that provided a short-term benefit, but created a significant long-term threat to U.S. vital interests. As with Reagan's predecessors, security assistance created unintended consequences and a long-term threat to the U.S. homeland that has yet to be eradicated.

Conversely, the security assistance programs to selected countries in the Persian Gulf and other parts of the Middle East were valuable in prosecuting and winning the 1991 Gulf War. Prior to FY 1990, over $15 billion in Foreign Military Sales construction projects in the Persian Gulf built critical infrastructure for strategic and tactical airlift, strategic sealift, and protected command and control capabilities. Additionally, this assistance enhanced compatibility of equipment and procedures among coalition forces. During the war, security assistance also provided emergency military and humanitarian supplies to Israel (Patriot missiles), Turkey (aircraft missiles and artillery), and Iraqi

---

[468] U.S. Department of State, *Fact Sheet: Usama bin Laden*, 1-2.
[469] U.S. Department of State, *Patterns of Global Terrorism 1999*, 7-8.
[470] George Tenet, Director of Central Intelligence, briefing to Senate Select Committee on Intelligence, "Worldwide Threat 2001: National Security in a Changing World," 7 February 2001, n.p.; on-line, Internet, 5 March 2000, available from http://www.cia.gov/cia/public_affairs/speeches/UNCLASWWT_02072001.html.

Kurds (humanitarian aid). In this case, security assistance proved invaluable in helping prepare for and prosecute the Gulf War.[471]

Reagan and Arms Control

Like his counter-terrorism and security assistance programs, Reagan only achieved limited success in the arms control arena. But he did make sweeping changes in our strategic nuclear forces policy that eventually brought about reductions in nuclear forces on both sides. He abandoned SALT, with its emphasis on ceilings on nuclear weapons, in favor of Strategic Arms Reduction Talks (START) in 1982. Instead of just seeking a ceiling on warheads, Reagan wanted to reduce the number of warheads on each side by about one-third, as well as the number of missiles on each side.[472] Reagan intended this proposal to reduce the Soviet ICBM threat against our own ICBMs. Acceptance of Reagan's proposal would have required the Soviets to reduce their ICBM warheads by 53 percent. Predictably, they were not receptive.

Then in March 1983, President Reagan changed the very nature of U.S. government deterrence policy. He abandoned the concept of MAD in favor of building a space-based missile defense system—the Strategic Defense Initiative (SDI). The goal of this program was to "intercept and destroy strategic ballistic missiles before they reached our own soil or that of our allies."[473] Predictably, the Soviets were opposed to the program, claiming that it violated the ABM treaty and was destabilizing. But their objections were more based on the fact that the Soviets had already developed an ABM capability in violation of the ABM treaty—and they wanted to maintain their advantage.

---

[471] U.S. Army Tank-automotive and Armaments Command (TACOM) Security Assistance Center, *The Bush Administration,* 21 November 2000, 1-2; on-line, Internet, 12 February 2001, available from http://www-acala1.ria.army.mil/tsac/bush.htm.
[472] Newhouse, 344-345.

There is ample evidence to indicate that for the Soviets, the ABM Treaty was merely a means to leave the U.S. vulnerable to strategic nuclear attack, while building up their own nuclear defense capability. By the 1980's, the U.S. had proof of significant Soviet violations, including:

    a.   Deployment of thousands of dual purpose anti-aircraft/anti-ballistic missile systems,[474] including the SA-5b (range 150 km, nuclear yield 25 kilotons),[475] SA-10 (range 5-90 km, nuclear yield unknown),[476] and the S-300V, which consists of two missiles: the dual-role SA-12a Gladiator anti-missile and anti-aircraft missile (range 6-75 km, 150 kg high explosive warhead) and SA-12b Giant anti-missile (range 13-100 km, 150 kg high explosive warhead),[477]—all in violation of the ABM treaty. According to former CIA analyst William Lee, by the time the Soviet Union collapsed, the Soviets had deployed over 10,000 SA-5/10 systems.[478] *Aviation Week and Space Technology* reported that during tests against tactical ballistic missiles, the S-300V system reportedly shot

---

[473] Foerster, 29.

[474] House, *U.S. National Missile Defense Policy and the Anti-Ballistic Missile Treaty: Hearings Before the Committee on Armed* Services, 106th Cong., 1st sess., 13 October 1999, 17; on-line, 13 January 2001, available from http://commdocs.house.gov/committees/security/has286000.000/has286000_0.HTM.

[475] Center for Defense Information, *Nuclear Weapons Database: Russian Federation Arsenal: Sea-Based Strategic Weapons: SA-5B Gammon SAM (S-200 Volga)*, 16 November 1998, n.p.; on-line, Internet, 13 January 2001, available from http://www.cdi.org/issues/nukef&f/database/rusnukes.html#sa5b.

[476] Center for Defense Information, *Nuclear Weapons Database: Russian Federation Arsenal: Sea-Based Strategic Weapons: SA-10 Grumble SAM (S-300)*, 16 November 1998, n.p.; on-line, Internet, 13 January 2001, available from http://www.cdi.org/issues/nukef&f/database/rusnukes.html#sa10

[477] Federation of American Scientists, *S-300V SA-12A GLADIATOR and SA-12B GIANT HQ-18*, 30 June 2000, 1; on-line, Internet, 13 January 2001, available from http://www.fas.org/nuke/guide/russia/airdef/s-300v.htm

[478] Lee, 13.

down over 60 600-km range tactical ballistic missiles with a 40-70 percent kill probability.[479]

b. Construction of the Krasnoyarsk ballistic missile detection and tracking radar facility 3700 km from Moscow (750 km from the Mongolian border), in violation of the ABM treaty's requirement that any such radars must be located within a 150-km radius of the national capital. In further violation, it was oriented not towards the Soviet border, but northeast across 4000 kilometers of Soviet territory.[480] According to then-Secretary of Defense Caspar Weinberger and then-Secretary of State George Schultz, this radar "closed the last remaining gap in Soviet ballistic missile detection coverage."[481]

Weinberger and Schultz were also concerned about the "growing Soviet network of large phased array ballistic missile detection and tracking radars," which would enable Soviet construction of a nation-wide ABM system, potentially within months, once the radars were complete.[482]

Reagan's SDI program, if it had been completed, would have fundamentally changed the strategic balance of power between the U.S. and Soviet Union by providing a powerful ABM capability for the U.S. However, both Congress and the scientific community had grave doubts about both the feasibility and cost of the program.[483] Research continued on SDI through the Reagan and Bush administrations, but with the

---

[479] Nikolay Novichkov and Michael Dornheim, "Russian SA-12, SA-10 On World ATBM Market," *Aviation Week and Space Technology* 146, no.9 (3 March 1997): 58.
[480] Weinberger and Schultz, n.p.
[481] Weinberger and Schultz, n.p.
[482] Weinberger and Schultz, n.p.
[483] Jordan, Taylor, and Korb, 79.

disintegration of the Soviet Union in 1991 and the signing of new arms control treaties with Russia, President Clinton abandoned SDI in 1993, and established the Ballistic Missile Defense Organization (BMDO) to research ground-based ABM systems.[484]

In terms of nuclear arms control, the Reagan administration's efforts did, however, build a foundation for later significant strategic arms reduction agreements. Although Reagan did not produce a new strategic nuclear forces treaty with Gorbachev or his predecessors, he did successfully negotiate the Treaty Between The United States Of America And The Union Of Soviet Socialist Republics On The Elimination Of Their Intermediate-Range And Shorter-Range Missiles (often dubbed the "INF Treaty") in 1987, which required destruction of all of each party's ground-launched cruise and ballistic missiles with ranges "between 500 and 5,500 kilometers, their launchers and associated support structures and support equipment within three years."[485]

Bush and Arms Control

When President Bush succeeded President Reagan in 1989, he and his Secretary of Defense, Richard Cheney, made an arms control treaty with the Soviets a top priority. The Bush administration built on the START negotiations of the Reagan era and managed to force the Soviets to drop their insistence that any strategic arms reduction treaty be linked to an agreement on missile defense.[486] In July 1991, Bush successfully negotiated a nuclear arms control treaty with the Soviets (START I) that would reduce

---

[484] Alex Tonello, *The Rise and Fall of the Strategic Defense Initiative*, 28 October 1997, 5; on-line, Internet, 11 January 2001, available from http://members.tripod.com/~atonello/sdi.htm.

[485] U.S. Department of State, *Treaty Between The United States Of America And The Union Of Soviet Socialist Republics On The Elimination Of Their Intermediate-Range And Shorter-Range Missiles*, 1987, n.p.; on-line, Internet, 13 March 2001, available from http://www.state.gov/www/global/arms/treaties/inf1.html.

[486] Federation of American Scientists, *Strategic Arms Reduction Treaty (START I) Chronology*, 6-7; on-line, Internet, 4 February 2001, available from http://www.fas.org/nuke/control/start1/chron.htm.

both U.S. and Russian nuclear warheads to 6000 each.[487] President Bush's ability to negotiate such a treaty was greatly enhanced by the fact that the Soviet Union was ready to fall apart at the time, and the U.S. had just won a major conventional war in the Persian Gulf.

Bush went further in decreasing nuclear tensions with the Soviets in September 1991 by unilaterally announcing that the U.S. would withdraw all of its land-based tactical nuclear weapons from overseas bases and sea-based tactical nuclear weapons from U.S. surface and sub-surface craft, stand down all strategic bombers from alert status, and immediately stand down all ICBMs scheduled for deactivation under START I, and end the mobile ICBM program. A week later, Gorbachev responded with a similar reduction.[488]

Arms reduction took another positive step forward with the November 1991 Nunn-Lugar Legislation, which authorized up to $400 million to help the Soviet Union destroy its stockpile of weapons of mass destruction (WMD) and to establish means to prevent proliferation of such weapons[489] (note: the Department of Defense defines WMD as "weapons that are capable of a high order of destruction and/or of being used in such a manner as to destroy large numbers of people").[490]

On December 25, 1991, the Soviet Union dissolved and became the Commonwealth of Independent States (CIS), and Boris Yeltsin became the new President of Russia. Yeltsin and Bush improved on the original START I treaty by signing the START II

---

[487] President, *National Security Strategy of the United States*, (Washington, D.C.: U.S. Government Printing Office, August 1991), 1, 5; on-line Internet, 1 February 2001, available from http://www.fas.org/man/docs/918015-nss.htm.
[488] Federation of American Scientists, *Strategic Arms Reduction Treaty (START I) Chronology*, 7.
[489] Federation of American Scientists, *Strategic Arms Reduction Treaty (START I) Chronology*, 7.
[490] DoD Dictionary of Military Terms, "Weapons of Mass Destruction," n.p.; on-line, Internet, 13 March 2001, available from http://www.dtic.mil/doctrine/jel/doddict/data/w/06784.html.

agreement to reduce U.S. and Russian nuclear warheads to between 3000-3500 each on ICBMs, SLBMs, and strategic bombers.[491]

The success of the Reagan and Bush arms reduction efforts was achieved in no small part from the military and economic strength from which Bush and Reagan negotiated. The U.S. emerged from the 1991 Gulf War the undisputed military power in the world. And by the time START I was signed, the Soviet Union was imploding—an August 1991 failed coup attempt against Gorbachev highlighted the weakness and internal divisions within the Soviet Union. By 25 December 1991, the Soviet Union no longer existed, and the Cold War was over.[492]

Military Force Reductions

With the end of the Soviet Union and victory in the 1991 Gulf War, the Bush administration continued the reductions in military force structure and budget it had begun before the 1991 Gulf War. General Colin Powell, then Chairman of the Joint Chiefs of Staff, knew the Democrat-controlled Congress would not support a Cold War-sized military, so he wanted to ensure that any force reductions were based on an overarching strategy in light of the new strategic environment. Even before the collapse of the Soviet Union, Powell had begun planning for a post-Cold War Base Force of about three-fourths the size of the Cold War military. Instead of a force sized to face the Warsaw Pact in the Fulda Gap, Powell envisioned a military sized to project power around the globe. Powell envisioned a military with capability to fight a major war in the Atlantic and Pacific, a deployable force to fight smaller contingencies, and a sufficient

---

[491] Federation of American Scientists, *Strategic Arms Reduction Treaty (START I) Chronology*, 7-10.
[492] _____, *CNN Cold War Episode 24: Conclusions*, 1-14; on-line, Internet, 4 February 2001, available from http://www-cgi.cnn.com/SPECIALS/cold.war/episodes/24/script.html.

nuclear force to deter nuclear attack.[493] Powell also wanted to ensure that U.S. Forces would be able to fight alone against "rogue" states—that is, "states [that] harbor aggressive intentions against their less powerful neighbors, oppose the spread of democracy, and are guilty of circumventing international norms against nuclear, biological, and chemical proliferation."[494]

Powell identified six "rogue" states that posed a threat to U.S. interests: Iran, Iraq, Syria, Libya, Cuba, and North Korea. To prevent any one or more nations from taking advantage of U.S. conflict with another state, he wanted to ensure that U.S. forces were sized large enough to defeat any two at once—hence the now famous two Major Regional Conflict (MRC) sized force. Powell also felt that we could not necessarily assume that any of our allies would help in a contingency, so our force structure had to be capable of prosecuting a two MRC strategy alone. General Powell felt this strategy required a military force structure about 75 percent of the size it was during the Cold War.[495]

In line with Powell's vision, total military personnel declined to 1.776 million in FY 1993 (down from 2.2 million in FY 1989). During the four-year Bush administration, the Army lost one-quarter of its strength (cut from 770,000 to 572,000), the Air Force lost 22 percent, the Navy 14 percent, and the Marines 9.7 percent of total strength.[496]

---

[493] Lawrence J. Korb, *U.S. National Defense Policy in the post-Cold War World,"* 14 June 2000, 1-2; on-line, Internet, 1 February 2001, available from http://www.foreignrelations.org/public/armstrade/korb_postcoldwar_paper.html.
[494] Korb, *U.S. National Defense Policy in the post-Cold War World,* 2-3.
[495] Korb, *U.S. National Defense Policy in the post-Cold War World,* 2-3.
[496] _____, "SecDef Histories—Richard Cheney," 5; on-line, Internet, 5 February 2001, available from www.defenselink.mil/specials/secdef_histories/bios/cheney.htm.

The Powell Doctrine

When Powell developed his concept for a reduced post-Cold War force structure, he envisioned the military would be used in a rational, logical manner to support U.S. defense interests. In fact, he developed a doctrine (now known as the Powell Doctrine) to try to ensure the U.S. military would be committed only under very specific circumstances. He specified three conditions under which the military should be used: a) to achieve clear and measurable political objectives, b) use overwhelming force decisively and quickly to accomplish the objective, and c) have a clear exit strategy. Powell was determined to prevent a repeat of Vietnam.[497]

The Cold War era ended with U.S. homeland defense capability at its zenith. Under Reagan's and Bush's defense policies, America achieved what the previous policy of containment could not—victory in the Cold War. With the implosion of the Soviet Union, coupled with a phenomenally one-sided military victory in the Gulf, America's national security was protected—not only because of its strong military, but also because of its credibility and will to protect its vital interests.

Analysis of American Homeland Defense: The Cold War

The advent of atomic weapons and the Cold War forever changed the U.S. government's policy on homeland defense. America could never again assume an isolationist stance in terms of protecting its homeland. Now, events in other countries around the world directly affected America's safety and security. This sea change in the global security environment profoundly impacted U.S. homeland security policy. America could no longer rely solely on its own technological and manpower resources to

---

[497] Korb, *U.S. National Defense Policy in the post-Cold War World*, 3.

defend itself. Homeland defense policy evolved into a much more complex phenomenon, relying on all of the instruments of our nation's power—economic, diplomatic, political, informational, and military—to protect our territory and national interests. In fact, with the nuclear age and the Cold War, the very concept of homeland defense changed. No longer was homeland defense a simply matter of protecting U.S. territory. Homeland defense was now inextricably tied to the strength of our alliances, the strength of our economy, access to global resources (especially oil) unchallenged access to global lines of communication, and our ability to defend not only our own homeland, but other strategically important countries threatened by internal instability, insurgencies, and hostile neighbors. This era also bought about a new challenge to American homeland defense—terrorism.

The U.S. government's response to these new homeland defense challenges was decidedly uneven in terms of success. Some of the same historic policy mistakes hindered our response to new homeland defense challenges:

    a.   With some exceptions, the government continually held force structure and defense policy hostage to imposed budget constraints. During both the Truman and Eisenhower eras, military force capability for handling regional and major conventional conflicts was severely hampered by large cuts in force structure. Especially during the Eisenhower era, defense policy (the New Look and New New Look) was at least partially determined by an emphasis on saving money at the expense of flexible, broad range capability. As a result, U.S. defense credibility suffered major setbacks, since we did not initially have the proper force structure to prosecute the Korean conflict, and our adversaries (correctly) did not believe we would resort to nuclear weapons to win the conflict. Similarly, during the Johnson, Nixon, Ford, and Carter eras, resolving growing U.S. economic problems and funding domestic programs were a key priority, and defense funding and policy suffered for it. As a result, America's international prestige and defense credibility suffered during and after Vietnam; America's military became a "hollow force" in the 1970's as America lost unchallenged control of sea lines of communication, the Soviets overtook America in nuclear weapons offensive capability, and America found itself unable to effectively project its military power to rescue the hostages in Iran. Schlesinger was forced to resign as Ford's Secretary of Defense because he would not countenance deep cuts in defense, despite obvious problems in U.S. military force capability. Carter's deep cuts in defense budgets left the U.S. bereft of capability to stop communist insurgent movements in Latin America or Africa, and unable to rescue the hostages held by Iran. The Reagan military buildup literally regenerated American defense capability, ensuring an overwhelming defeat of one of the world's largest armies during the 1991 Gulf War. Continuing the Reagan administration's policy of sizing military forces to the current threat environment, the Bush administration's force reductions were not based on an artificially imposed budget ceiling; rather, they were the result of General Powell's study of current threats and force structure during the waning years of the Cold War and after the Cold War.

b. National security planning became more centralized during the Truman administration, but the quality of policy planning did not improve until the Reagan administration. Congress passed the National Security Act of 1947 as a means to provide unified command and control of national security policy. Although this legislation did manage to centralize defense policy planning, it did not necessarily improve the quality of planning. Eisenhower's New Look and New New Look created a gross force imbalance while weakening our capability to prosecute conventional conflicts. And although McNamara and Kennedy rebuilt conventional force structure to fight across a spectrum of conflict, they never developed a clear policy for use of these forces, nor guidance on legitimate use of force in the national interest. Lack of coherent policy direction (for example, which countries were considered a great enough threat to our vital interests to warrant use of military force) brought about the failed Bay of Pigs invasion and resultant Cuban Missile Crisis. Additionally, McNamara's fallacious and intellectually bankrupt concept of MAD ignored the reality of Soviet defense efforts and counterforce targeting, severely constrained nuclear force employment and targeting options, and left America defenseless against Soviet missiles. The Nixon administration did little to improve the situation with the SALT I treaty, which ensured viability of the Soviet's counterforce nuclear strategy, and the ABM treaty which enabled the Soviets to develop ABM capability while ensuring the U.S. had no such defense. Carter's national security policy left the U.S. in even worse shape than his predecessors. Carter had no clear vision for U.S. national security or capability; instead, he emphasized arms control and lower defense budgets without a clear understanding of the strategic ramifications of his decisions, given the geo-political threat environment at the time. Carter's naiveté concerning Soviet intentions contributed to his policy of giving defense capability a lower priority than domestic programs. His national security team left the U.S. vulnerable and potentially unable to effectively retaliate against a Soviet first strike. Furthermore, under his administration, the U.S. lost absolute control of sea lines of communication, and ensured the Soviets had the time to build up their conventional forces to the point where NATO was now susceptible to attack from the Warsaw Pact. President Reagan, however, did have a clear goal in his homeland defense policy: "we win, they lose."[498] He achieved this through a massive military buildup and concomitant change in military strategy to a more forward-based, offensive stance. He changed U.S. strategic nuclear policy by abandoning SALT in favor of actual reductions in nuclear arms on both sides. Although the Soviets were not receptive his proposals on reducing ICBMs, he did manage to get the Soviets to agree to end deployment of intermediate range nuclear missiles. But his most sweeping achievement in strategic policy was abandonment of MAD as a basis for U.S. strategic nuclear policy. Instead, he initiated a program to develop space-based missile defenses. Although his program was eventually abandoned after the Cold War, the program was enough of a threat to the strategic balance of power that the Soviets did end

---

[498] Fox, 5.

up signing the first START treaty with the Bush administration. The Bush administration continued a positive, results-oriented strategic arms reduction policy with the new Commonwealth of Independent States (CIS) by unilaterally de-alerting ICBMs scheduled for deactivation and strategic bombers, and withdrawing tactical nuclear weapons from overseas bases and naval assets (an action which obligated Gorbachev to respond in kind) and by negotiating a further reduction in nuclear arms through START II. The Nunn-Lugar legislation to aid the former Soviet Union in destroying its WMD stockpile was another important means of reducing the threat of accidental or intentional WMD attack.

c. Civil defense programs, designed to protect the populace from nuclear fallout, were a complete failure. Basically, civil defense programs were a propaganda ploy to make the populace believe that nuclear war was survivable. Even before McNamara's tenure, Congress never provided the funding necessary to develop a robust shelter program for targeted cities. Eisenhower's city evacuation plan was unrealistic, given the numbers of potential evacuees and limitations on warning time and evacuation routes. McNamara, believing that a robust civil defense program would increase the risk of attack, oversaw the demise of what was left of civil defense programs. Given the destructive capability of WMD, the only real means to protect the U.S. populace is to take all necessary domestic and international measures to ensure a WMD attack never occurs in the U.S.

d. Security assistance programs proved to be a double-edged sword in terms of Cold War homeland defense. Planned and administered properly, with specific goals and implemented for short duration, they enhanced U.S. security by checking Soviet expansionism and tipping the strategic balance of power in endangered regions in favor of the U.S. The Truman-era Marshall Plan and aid to Greece and Turkey were both examples of successful assistance programs that enhanced U.S. homeland defense capability by strengthening the economic and political stability in regions threatened by internal instability and Soviet expansionism. Some subsequent security assistance programs were not as effective, and in fact, were deleterious to homeland defense capability. Lack of clearly defined, achievable objectives, lack of a pre-determined end point for security assistance, failure to comprehend the limits of what security assistance can achieve, and most importantly, failure to analyze and consider unintended consequences all contributed to disastrous effects of security assistance in Vietnam, Nicaragua, and Afghanistan. Conversely, security assistance programs in the Persian Gulf and select Middle Eastern countries provided needed infrastructure improvements and weapons systems that aided U.S. and coalition efforts in the Gulf War. Based on cases mentioned here, security assistance was beneficial to homeland defense efforts only when such assistance was given for a finite period of time, with clear, specific, limited goals. Security assistance programs detracted from homeland defense efforts when programs were implemented without considering unintended consequences, without a clear, achievable objective, and without a time limit for such assistance.

e. Collective security arrangements became an important part of U.S. defense policy to prevent further Soviet expansion and project U.S. political and military power in strategically critical regions. Although some did not survive very long (such as SEATO), NATO became the linchpin of U.S. and European security; NATO thwarted further Soviet expansion in Europe, and survived to become an important collective defense organization in the post-Cold War era.

f. Counter-terrorism programs during the Cold War did not achieve desired results. Lack of a sustained, consistent, effective policy and lack of robust HUMINT capability ensured continued attacks on Americans overseas during the Cold War.

Chapter 5

American Homeland Defense: The Clinton Era

It must be obvious, therefore, that periods of tranquility are rich in sources of friction between soldiers and statesmen, since the latter are forever trying to find ways of saving money, while the former are constantly urging increased expenditure. It does, of course, occasionally happen that a lesson recently learned, or an immediate threat, compels them to agree.

—Charles de Gaulle
The Edge of the Sword

Background

When William Clinton entered the White House in 1993, he faced a national security environment very different from his Cold War-era predecessors. During the Cold War, U.S. national security policy and objectives focused first on containment of communism, then during the Reagan era, defeat of communism as a major geo-political force. The Soviet Union and its allies were the primary threats to U.S. national interests, so national security policies and objectives were focused on stopping the spread of Soviet influence and on preventing a major conventional or nuclear conflict with the Soviet Union.[499] The Bush Presidency foretold the new threats to security that America would face over the next decade. Regional threats (such as Iraq and Iran), and managing the economic and political fallout in the wake of dissolution of the Soviet Union and its East European satellites became key national security issues.[500]

During the 8 years of the Clinton Administration, the U.S. faced an even more diverse threat environment—one in which physical distance and territorial boundaries no longer provided a modicum of protection. The threat of nuclear war with Russia was no longer a priority homeland defense issue, but security and command and control over Russia's nuclear stockpile became a priority concern. Proliferation of WMD and ballistic missile technology among rogue states, spreading ethnic and regional conflicts, and mass casualty terrorist attacks against Americans overseas and in the U.S. all became priority threats to U.S. homeland security.[501] Based on the increasingly diverse threats to U.S.

---

[499] President, *National Security Decision Directive Number 32, U.S. National Security Strategy* (20 May 1982), 1-2.
[500] U.S. Army Tank-automotive and Armaments Command (TACOM) Security Assistance Center, *The Bush Administration,* 1.
[501] President, *A National Security Strategy of Engagement and Enlargement 1996,* (Washington, D.C.: Government Printing Office, February 1996), ii-iii; on-line, Internet, 21 February 2001, available from http://www.fas.org/spp/military/docops/national/1996stra.htm.

territory and vital interests, the requirements for a viable homeland defense policy now encompassed nearly all aspects of U.S. international and domestic policy. Essentially, homeland defense became synonymous with national security.

Terrorism: The Greatest Challenge to U.S. Homeland Defense

Of all the new threats to U.S. homeland security evidenced during the Clinton administration, terrorism became the most immediate, urgent threat to the U.S., thanks to proliferation of WMD technologies and terrorist proclivity to enact mass casualty attacks. Terrorist groups now possessed both the capability and intent to execute chemical and biological attacks. In March 1995, Aum Shinrikyo, a Japanese apocalyptic religious cult, used sarin gas in a Tokyo subway attack, killing 12 people and injuring several thousand. Prior to the attack, this group experimented with both anthrax and botulinum toxin.[502] In 1995, a right-wing U.S. anti-tax extremist group, the Patriot's Council, was convicted of attempting to use ricin, a poison extracted from castor bean, in a attempt to kill local Minnesota officials. Had the ricin been effectively used, it could have killed over 100 people.[503]

These precedent-setting attacks may make other terrorist groups more inclined to use chemical and biological weapons in the future. In fact, many known terrorist organizations are interested in obtaining unconventional weapons. Usama bin Laden, leader of al-Qa'ida (responsible for the August 1998 U.S. Embassy bombings and suspected mastermind of the October 2000 U.S.S. Cole bombing), declared acquisition of

---

[502] Kyle B. Olson, "Aum Shinrikyo: Once and Future Threat?" *Emerging Infectious Diseases*, 5, no. 4 (July-August 2000), n.p.; on-line, Internet, 23 February 2001, available from www.cdc.gov/ncidod/EID/vol5no4/olson.htm.
[503] Federal Bureau of Investigation, Counter-terrorism Threat Assessment and Warning Unit, National Security Division, *Terrorism in the United States 1996*, 24; on-line, Internet, 23 February 2001, available from http://www.fbi.gov/library/terror/terroris.pdf.

these weapons a "religious duty."[504] The United States District Court, Southern District of New York indictment against Usama bin Laden and his co-conspirators in the 1998 U.S. Embassy bombings implicated bin Laden in attempts to obtain both nuclear and chemical weapons.[505]

In addition to demonstrated use of and interest in chemical, nuclear, and biological weapons, terrorist groups showed a proclivity for attacks specifically designed to kill as many people as possible. The 1993 World Trade Center bombing (6 killed, over 1000 injured) was actually intended to destroy both towers of the World Trade Center, which would have caused tens of thousands of casualties.[506] The 1995 bombing of the Alfred P. Murrah Federal building in Oklahoma City (168 killed),[507] the 1998 U.S. embassy bombings in Kenya and Tanzania (391 killed, over 5000 wounded),[508] the 1996 Khobar Towers bombing in Dhahran, Saudi Arabia (19 U.S. citizens killed, over 500 persons wounded),[509] and the suicide bombing of the U.S.S. Cole (17 killed)[510] all showed a trend among terrorists to produce mass casualties.

The most dangerous of the terrorist groups to emerge was Usama bin Laden and his Al Qa'ida terrorist organization. This group maintained a steady, effective campaign of

---

[504] Tenet, "Worldwide Threat 2001: National Security in a Changing World," n.p.
[505] Mary Jo White, United States Attorney, "Indictment S(9) 98 Cr. 1023 (LBS) United States of America versus Usama bin Laden [and other associates]," 19; on-line, Internet, 25 February 2001, available from http://www.fbi.gov/majcases/eastafrica/indictment.pdf.
[506] National Commission on Terrorism, *Countering the Changing Threat of International Terrorism*, 5 June 2000, 7; on-line, Internet, 3 February 2001, available from http://www.fas.org/irp/threat/commission.html.
[507] _____, "Oklahoma Bombing," *Washington Post*, (no date) 1997, 1; on-line, Internet, 23 February 2001, available from http://washingtonpost.com/wp-srv/national/longterm/oklahoma/oklahoma.htm.
[508] State Department, Office of the Coordinator for Counter-terrorism, *Patterns of Global Terrorism 1998*, April 1999, n.p.; on-line, Internet, 23 February 2001, available from http://www.state.gov/www/global/terrorism/1998Report/1998index.html.
[509] U.S. Department of State, Office of the Coordinator for Counter-terrorism, *1996 Patterns of Global Terrorism Report* (Washington, D.C.: U.S. Government Printing Office, 1997), n.p.; on-line, Internet, 23 February 2001, available from http://www.state.gov/www/global/terrorism/1996Report/middle.html.
[510] Tenet, "Worldwide Threat 2001: National Security in a Changing World," n.p.

terror against the U.S. and its vital interests throughout the Clinton administration. Bin Laden was linked to several attempted terrorist attacks, including plots to kill the Pope in 1994 and President Clinton in 1995 during their respective visits to the Philippines, as well plots to bomb a dozen U.S. trans-Pacific flights in 1995.[511] On 23 February 1998, bin Laden issued a "fatwa" (a legal opinion or decree from an Islamic religious leader)[512] to kill Americans and their allies, in any country where it is possible, to liberate Mecca and to get U.S. armies out of all Islamic lands.[513] His organization was responsible for the 7 August 1998 U.S. Embassy bombings in Kenya and Tanzania,[514] and evidence is mounting that he was responsible for the 12 October 2000 bombing of the U.S.S. Cole.[515]

The Clinton administration's response to these terrorist campaigns was not effective. Although U.S. government agencies successfully prevented some major Islamic extremist terrorist attacks on U.S. soil, including the planned bombings of California's Disneyland and Seattle Space Needle,[516] the United Nations complex, and Lincoln and Holland tunnels between New Jersey and New York City,[517] terrorists continued to stage major

---

[511] U.S. Department of State, "Background Information on Terrorist Groups," *Patterns of Global Terrorism 1998*, (Washington, D.C.: U.S. Government Printing Office, 1999) n.p.; on-line, Internet, 25 February 2001, available from http://www.state.gov/www/global/terrorism/1998Report/appb.html.
[512] Merriam-Webster's Collegiate Dictionary, "Fatwa" on-line, Internet, 25 February 2001, available from http://www.m-w.com/cgi-bin/dictionary?book=Dictionary&va=fatwa.
[513] _____, "Text of Fatwa Urging Jihad Against Americans," *Al-Quds al-'Arabi* (in Arabic), 23 February 1998, 3; on-line, Internet, 25 February 2001, available from http://www.emergency.com/bladen98.htm.
[514] State Department, "Background Information on Terrorist Groups," *Patterns of Global Terrorism 1998*, n.p.
[515] Rym Brahimi, Peter Bergen, and David Ensore, "U.S. Finds Link Between bin Laden and Cole Bombing," *Cable News Network*, 7 December 2000, on-line, Internet, 25 February 2001, available from http://www.cnn.com/2000/US/12/07/cole.suspect/.
[516] Richard Sale, "Terrorists Targeted Disneyland, Space Needle," *United Press International*, 20 February 2001.
[517] Andrew C. McCarthy, "Prosecuting the New York Sheikh," *Middle East Quarterly* (March 1997), n.p.; on-line, Internet, 25 February 2001, available from http://www.ict.org.il/articles/articledet.cfm?articleid=95.

attacks against U.S. citizens and our allies throughout the 1990's. The August 1998 retaliatory U.S. missile attacks on bin Laden's terrorist camps in Afghanistan did some damage to terrorist training camps but missed bin Laden;[518] he is still financing and directing terrorist activities from Afghanistan.[519]

These attacks were just as ineffective as Reagan's one-time strike on Libya in 1986. Destruction of a state's or organization's terrorist capability requires a sustained campaign using all instruments of national power; the campaign cannot stop until the reason for the campaign no longer exists. The state or organization must either renounce terrorism or face destruction.

Furthermore, the Khobar Towers bombing incident remains an open wound. Saudi officials named Iranian-backed Shiite extremists as the perpetrators of the bombing. The alleged ringleader of the bombers, Ahmad Mughassil (a member of the Saudi Hezbollah), fled to Iran with some of his followers within hours of the bombing. Iran continues to deny the bombers are in Iran. The Clinton administration reportedly refused to press the issue, allegedly because once the U.S. public knew of Iran's complicity, Clinton would have been forced to act against Iran—and he didn't want to undermine a potential rapprochement with Iran's new President Khatami.[520]

Another terrorism policy failure was the Clinton administration's response to the Iraq plot to kill former President George W. Bush. Kuwaiti intelligence uncovered an Iraqi plot to kill former President Bush with a car bomb during his visit to Kuwait in April

---

[518] Jamie McIntyre, "Moderate to Severe Damage Seen at Suspected bin Laden Camps," *Cable News Network*, 13 January 1999, on-line, Internet, 25 February 2001, available from http://www.cnn.com/WORLD/asiapcf/9901/13/afghan.damage.photos/.
[519] _____, "Bin Laden Makes a Move," *CBS News*, 13 November 2000, n.p.; on-line, Internet, 25 February 2001, available from http://cbsnews.com/now/story/0,1597,206750-412,00.shtml.
[520] Kenneth R. Timmerman, "Is Iran-Saudi Détente Underway?" *Wall Street Journal Europe*, 20 May 1999, n.p.; on-line, Internet, 26 February 2001, available from http://www.iran.org/tib/krt/wsje_990520.htm.

1993. Of the 14 suspects captured by Kuwaiti officials, several worked for Iraqi intelligence.[521] Forensic evidence clearly showed Iraqi intelligence involvement—the bomb had the same remote control firing device and other key components that Iraqi intelligence typically used. President Clinton ordered a cruise-missile attack on Iraqi intelligence headquarters in response to the incident.[522] The raid was deliberately conducted at night, so none of the people responsible for the plot against former President Bush were in at the time.[523] As a result, Iraq continued its campaign of state-sponsored terror in the years since, focusing on anti-government opposition within and outside of Iraq, and hosting various terrorist groups such as the Abu Nidal Organization. Press reports indicate that Iraqi intelligence plotted to bomb Radio Free Europe in Prague in 1998, but the plot was foiled when it became public.[524]

It is clear that the Clinton administration's terrorism policy was a failure. Of the major international terrorist threats to U.S. interests in the 1990's, President Clinton's administration neither nullified nor significantly weakened any of them. Like his predecessors, he failed to develop a steady, comprehensive, long-term policy to fight international terrorists.

Ballistic Missile Proliferation

---

[521] U.S. Department of State, Office of the Coordinator for Counter-terrorism, *Patterns of Global Terrorism 1993* (Washington, D.C.: U.S. Government Printing Office, 1994), n.p.; on-line, Internet, 26 February 2001, available from http://www.hri.org/docs/USSD-Terror/93/statespon.html#Iraq.

[522] W. Michael Reisman, "The Raid on Baghdad: Some Reflections on its Lawfulness and Implications," *American Journal of International Law*, 5, no. 1 (no date) n.p.; on-line, Internet, 26 February 2001, available from http://www.ejil.org/journal/Vol5/No1/art11.html.

[523] DoD News Briefing, "Pentagon Details U.S. Missile Attack on Iraq," 26 June 1993, on-line, Internet, 26 February 2001, available from http://www.fas.org/man/dod-101/ops/docs/dod_930626.htm.

[524] U.S. Department of State, Office of the Coordinator for Counter-terrorism, *Patterns of Global Terrorism 1999* (Washington, D.C. U.S. Government Printing Office, 2000), n.p.; on-line, Internet, 26 February 2001, available from http://www.state.gov/www/global/terrorism/1999report/sponsor.html#Iraq.

In addition to terrorism, proliferation of ballistic missiles (with WMD payload capability) among hostile states became a significant threat to the U.S. homeland and vital interests. Although the President's own 1996 National Security Strategy stated that, "Weapons of mass destruction -- nuclear, biological and chemical -- along with their associated delivery systems, pose a major threat to our security and that of our allies and other friendly nations,"[525] President Clinton vetoed the 1996 defense authorization bill that called for deployment of a U.S. missile defense by 2003, saying it "cannot be accommodated within the terms of the existing ABM Treaty," choosing to discount the possibility of new near-term ballistic missile threats.[526]

House National Security Committee Chairman Floyd Spence disagreed with both the President's decision not to deploy a missile defense and his reasons for it. So he included a provision in the 1997 Defense Authorization Act that established an independent commission to review current assessments of ballistic missile threats.[527] The resulting commission, chaired by present Secretary of Defense Donald Rumsfeld, was dubbed "Commission to Assess the Ballistic Missile Threat to the United States," better known as the "Rumsfeld Commission."[528] The unclassified Executive Summary of the July 1998 Rumsfeld Commission report stated that, in addition to Russia's and China's current ballistic missile capabilities, several states could "inflict major destruction on the U.S. within about five years of a decision to acquire such a capability (10 years in the case of

---

[525] President, *A National Security Strategy of Engagement and Enlargement 1996*, n.p.
[526] Floyd Spence, Chairman, House National Security Committee, "Unveiling the Ballistic Missile Threat: The Ramifications of the Rumsfeld Report," *National Security Report*, 2, issue 4 (August 1998) 3; on-line, Internet, 27 February 2001, available from http://www.house.gov/hasc/Publications/105thCongress/NSRs/nsr2-4rumsfeldreport.pdf.
[527] Spence, 1-3.
[528] Spence, 3.

Iraq)."[529] Iran, Iraq, and North Korea were named as countries of major concern, since their targets include not only the U.S. homeland, but also deployed U.S. forces and key allies of the U.S.[530]

The Rumsfeld Commission's assessment turned out to be correct. One week after the 15 July 1998 publication of the Rumsfeld report (which also concluded that Iran could flight test its 1300 km range Shahab-3 MRBM at any time and deploy it soon thereafter),[531] Iran performed its first flight test of the Shahab-3.[532] The Rumsfeld Commission also assessed that Iran has the capability to "demonstrate an ICBM-range ballistic missile…within five years of a decision to proceed…"[533] Furthermore, the Rumsfeld Commission assessed that Iran is developing weapons of mass destruction and intends to produce nuclear weapons as soon as possible.[534] This assessment is especially disturbing in light of recent reports that Russia is assisting Iran in development of a nuclear reactor and is supplying "over $7 billion worth of weapons [to Iran] over the next few years…"[535]

The Rumsfeld Commission was also correct in its assessment of North Korea's intentions and capability. On 31 August 1998, just 6 weeks after publication of the Rumsfeld Commission report, North Korea launched a multistage Taepo Dong 1 MRBM

---

[529] Donald H. Rumsfeld et al., *Executive Summary of the Report of the Commission to Assess the Ballistic Missile Threat to the United States*, 104th Cong., 15 July 1998, 2-3; on-line, Internet, 26 February 2001, available from http://www.fas.org/irp/threat/bm-threat.htm.
[530] Rumsfeld et al, 6.
[531] Rumsfeld et al., 7.
[532] Spence, 3.
[533] Rumsfeld et al, 7.
[534] Rumsfeld et al, 7.
[535] Anwar Faruqi, "Russia Arming An Iran In Disputes With Almost All Its Neighbors," *Associated Press*, 20 March 2001, n.p.; on-line, Internet, 20 March 2001, available from http://ebird/dtic.mil/Mr2001/e20010320arming.htm.

over Japan in a failed attempt to orbit a satellite[536]—demonstrating future potential to use a ballistic missile to hit U.S. territories in the Pacific, Alaska, and Hawaii.[537]

North Korea's indigenous ballistic missile program is only part of the threat this country poses—it also has been exporting ballistic missile technology to other nations, including Iran and Pakistan.[538] According to February 2000 testimony of Director of Central Intelligence George Tenet, "proliferation of medium-range ballistic missiles (MRBMs)—driven primarily by North Korean No Dong sales—is significantly altering the strategic balances in the Middle East and Asia."[539]

Russia and China are also responsible for much the proliferation of missile technology into the Middle East and Asia. According the to the Rumsfeld Commission report,

> Russia poses a threat to the U.S. as a major exporter of enabling technologies, including ballistic missile technologies, to countries hostile to the United States. In particular, Russian assistance has greatly accelerated Iran's ballistic missile program...China also poses a threat to the U.S. as a significant proliferator of ballistic missiles, weapons of mass destruction, and enabling technologies...It has supplied Pakistan with a design for a nuclear weapon...[and] has even transferred complete ballistic missile systems to Saudi Arabia...and Pakistan...[540]

---

[536] Robert D. Walpole, National Intelligence Officer for Strategic and Nuclear Programs, "North Korea's Taepo Dong Launch and Some Implications on the Ballistic Missile Threat to the United States," speech, Center for Strategic and International Studies, Washington D.C., 8 December 1998; on-line, Internet, 26 February 2001, available from http://www.cia.gov/cia/public_affairs/speeches/archives/1998/walpole_speech_120898.html.

[537] Kim R. Holmes et al, "Preface," *Defending America: A Plan to Meet the Urgent Missile Threat*, Heritage Foundation Report, (Washington, D.C.: Heritage Foundation Commission on Missile Defense, March 1999), 1-2; on-line, Internet, 27 February 2001, available from http://www.heritage.org/missile_defense/preface.html.

[538] Rumsfeld et al, 7.

[539] George J. Tenet, Director of Central Intelligence, "The Worldwide Threat in 2000: Global Realities on Our National Security," statement to Senate Select Committee on Intelligence, 2 February 2000, 1-4; on-line, Internet, 26 February 2001, available from http://www.usinfo.state.gov/topical/pol/terror/00020201.htm.

[540] Rumsfeld et al, 5.

Despite overwhelming evidence of China's role in exporting ballistic missile technology and Chinese theft of U.S. nuclear weapons secrets, the Clinton administration did nothing to sanction China or prevent its further missile technology exports—in fact, the administration pushed through Most Favored Nation (MFN) trade status for China in the summer of 2000.[541]

Iraq is another country whose WMD and ballistic missile programs pose a direct threat to U.S. forces and vital interests. Iraq already demonstrated capability for WMD attacks during the Iran-Iraq War, when the Iraqi Army used chemical weapons against both the Iranian Army[542] and Kurdish civilians.[543] During the Gulf War, our attempt to destroy Iraq's WMD program failed—most of Iraq's biological and chemical weapons facilities survived the end of the 1991 Gulf War.[544] After the 1991 Gulf War, Iraq declared it had produced and weaponized anthrax, botulinum toxin, and aflatoxin—but claimed it had since destroyed them all. UN inspectors, however, found evidence of much more in terms of chemical and biological weapons capability than Iraq had declared.[545]

After Iraq barred UN weapons inspectors in 1998, the U.S. and Britain conducted a 4-day campaign of air and missile strikes in December 1998 against Iraq's political-military infrastructure and suspected WMD facilities (Operation Desert Fox). Although

---

[541] Robert Kagan, "The Clinton Legacy Abroad," *Weekly Standard*, 15 January 2001, 25; on-line, Internet, 16 January 2001, available from http://dailyread/esup/tues/s20010116legacy.htm.
[542] Julian Perry Robinson and Jozef Goldblat, *Chemical Warfare in the Iraq-Iran War*, Stockholm International Peace Research Institute (SIPRI) Fact Sheet (Stockholm, Sweden: SIPRI, 1984), 1-2; on-line, Internet, 13 March 2001, available from http://projects.sipri.se/cbw/research/factsheet-1984.html.
[543] Senate Resolution 408, 100th Cong., 2nd sess., 24 June 1988, on-line, Internet, 26 February 2001, available from http://www.senate.gov/~rpc/rva/1002/1002201.htm.
[544] Anthony H. Cordesman, "The Military Effectiveness of Desert Fox: A Warning About the Limits of the Revolution in Military Affairs and Joint Vision 2010 (working draft)," 26 December 1998, 5; on-line, Internet, 26 February 2001, available from http://www.csis.org/stratassessment/reports/effectiveDesertFox.pdf.
[545] Office of the Assistant Secretary of Defense (Public Affairs), "Iraq's Chemical and Biological Weapons Capability," briefing 14 November 1997, on-line Internet, 26 February 2001, available from http://www.defenselink.mil/news/Nov1997/x11171997_x114iraq.html.

the U.S. and Britain did damage Iraq's military and WMD infrastructure during the campaign, the campaign did not solve the problems of forcing Iraq to let inspectors back in or of ending Iraq's WMD program. In fact, the campaign garnered sympathy for Hussein among some members of the UN and enhanced Hussein's stature in the Arab world for standing up to the U.S.[546]

According to testimony of George Tenet, Director of Central Intelligence, since the end of WMD inspections in Iraq, it is difficult to assess the status of Iraq's WMD programs; however, there is mounting concern that repair of facilities damaged during Desert Fox could be associated with WMD programs, and the CIA assumes Saddam Hussein continues to give WMD capability a high priority.[547]

The upshot of Clinton's policy regarding Iraq's WMD program is that for the past 3 years since Desert Fox, Hussein has not allowed UN weapons inspectors back into Iraq, and international support for sanctions against Hussein's intransigence is quickly crumbling.[548] In terms of U.S. homeland defense, Clinton's policy was a failure.

In essence, proliferation of ballistic missiles and associated WMD technology among states hostile to the U.S. and its allies now can impede U.S. power projection, threaten U.S. forces, vital interests, and allies overseas.[549] It is only a matter of time before these states can directly threaten the U.S. homeland.

---

[546] Cordesman, "The Military Effectiveness of Desert Fox: A Warning About the Limits of the Revolution in Military Affairs and Joint Vision 2010 (working draft)," 30-31.
[547] George J. Tenet, Director of Central Intelligence, "The Worldwide Threat in 2000: Global Realities on Our National Security," 9.
[548] Kagan, 25.
[549] George J. Tenet, Director of Central Intelligence, "The Worldwide Threat in 2000: Global Realities on Our National Security," 1-4.

Russia's Command and Control of Nuclear Weapons

Another threat to the U.S. homeland that developed during the past decade was Russia's weakened control over its nuclear weapons. Russia's severe economic problems have rendered it unable to properly maintain a large standing military. Concerns center around accidental or unauthorized launch due to faulty command and control, and possible theft, sale, or loss of nuclear weapons, and associated materials. A November 2000 Congressional Research Service Issue Brief for Congress reported that Russian Defense Minister Rodionov pointed out in 1997 that "he feared a loss of control over Russian strategic nuclear forces in the future if additional funding were not available to maintain and modernize the communications links in the nuclear command and control structure."[550] Inadequate funding for pay and training and low morale among Russian Strategic Rocket Forces personnel, inadequate security and poor record-keeping at Russian nuclear weapons storage facilities, and concerns about possible theft of nuclear materials due to deteriorating economic conditions also contribute to potential loss or unauthorized/accidental use of Russian nuclear weapons.[551]

Clinton and National Missile Defense

Unfortunately, the Clinton administration did little to ameliorate the growing threat from ballistic missiles and WMD technology or Russia's deteriorating command and control over its nuclear forces. He cancelled former President Bush's Global Protection Against Limited Strikes (GPALS) missile defense program, cut National Missile Defense

---

[550] Amy F. Woolf, "IB98038: Nuclear Weapons in Russia: Safety, Security, and Control Issues," *Congressional Research Service Issue Brief for Congress*, 21 November 2000, 3-4; on-line, Internet, 28 February 2001, available from http://www.cnie.org/nle/inter-64.html#_1_3.
[551] Woolf, 4-5.

(NMD) funding by 80 percent in 1993,[552] multilateralized the antiquated and legally defunct ABM Treaty by recognizing Russia, Belarus, Kazakhstan, and Ukraine as the successors to the Soviet Union for treaty purposes,[553] and vetoed the 1996 Defense Authorization bill which required deployment of an NMD system by 2003.[554] Even after the August 1998 North Korean Taepo Dong launch, and after he signed the 1999 National Missile Defense Act (which committed the U.S. to develop and field an NMD system "as soon as technologically feasible"), President Clinton deferred a decision to deploy a national missile defense to the next President.[555] Now, the United States is in a catch-up mode to try to develop an effective national missile defense before our enemies develop the capability to launch an effective attack on the U.S. homeland.

Defense Against Conventional Threats

Clinton's defense policy toward more conventional threats was not successful in improving homeland defense capability, either. His *National Security Strategy of Engagement and Enlargement* emphasized engagement abroad to deter and resolve conflicts.[556] When translated into action, it eroded U.S. military readiness by using up shrinking military equipment and personnel resources on peripheral conflicts.

The problem started in 1993, when President Clinton doubled the original 5-year budget cuts (initiated by former President Bush and his Secretary of Defense Dick

---

[552] Baker Spring, "Clinton's Failed Missile Defense Policy: A Legacy of Missed Opportunities," *The Heritage Foundation Backgrounder* no. 1396 (21 September 2000): 1-4; on-line, Internet, 21 February 2001, available from http://www.heritage.org/library/backgrounder/bg1396.html.
[553] Kim R. Holmes et al, "Chapter 2 The ABM Treaty and Intentional Vulnerability," *Defending America : A Plan to Meet the Urgent Missile Threat*, Heritage Foundation Report, (Washington, D.C.: Heritage Foundation Commission on Missile Defense, March 1999), 2; on-line, Internet, 27 February 2001, available from http://www.heritage.org/missile_defense/chapter2.html.
[554] Spring, 3.
[555] Spring, 1-2.
[556] President, *A National Security Strategy of Engagement and Enlargement 1996*, i-iii.

Cheney) to $128 billion. The U.S. Navy shrank from 443 ships in 1993 to 316.[557] The Army shrank from 14 to 10 divisions.[558] Overall, the U.S. military shrank in the 1990's by 40 percent.[559] Despite the budget cuts, deployments increased from Cold War era levels by 300 percent. During the entire Cold War, the U.S. engaged in only 16 small-scale contingencies. But by 1999, U.S. troops had deployed in 48 such operations, including conflicts in Iraq and Kosovo.[560] Predictably, equipment began to wear out, readiness suffered, and people began to leave the military.

In testimony to the Senate Armed Services Committee on 26 September 2000, the Joint Chiefs painted a gloomy picture of military readiness. They reported cannibalization of parts to keep aircraft flying, lack of spare parts to keep weapons in working order, military personnel working back-to-back deployments, and people working extra hours to keep old equipment functioning. The Service Chiefs reported multi-billion-dollar lists of requirements that Congress did not fund. The services were forced to rob modernization accounts to keep current equipment working. Two years ago, readiness deficiencies were most apparent in non-deployed forces. Now readiness deficiencies were showing up in deployed forces. Shortfalls in key support systems, such as strategic lift, tankers, and intelligence, surveillance and reconnaissance (ISR) assets were affecting our ability to fight. Due to the high operations tempo, the U.S. had "leveraged readiness on the backs of soldiers and their families;" as a result, experienced people were leaving the military.

---

[557] Rowan Scarborough, "Readiness of Armed Forces is Not Improving; Clinton Action on Pentagon Cuts Seen as Cause of Problem," *The Washington Times*, 28 August 2000, n.p.
[558] Douglas Austin, "Can America Fight Two Wars At Once? That's the Plan, but Experts Doubt It," *Investor's Business Daily*, 28 August 2000, n.p.
[559] Daniel Goure and Jeffrey M. Ranney, *Averting the Defense Train Wreck in the New Millennium* (Washington, D.C.: The CSIS Press, 1999), xi.
[560] Scarborough, n.p.

Due to these factors, the Joint Chiefs assessed the overall risk for executing a 1 Major Theater War (MTW) scenario as moderate, and a 2 MTW scenario as high.[561]

The Clinton administration and Congress failed, as so many previous administrations before, to match resources to strategy and commitments. As a result, U.S. conventional military defense capability eroded, and United States vital national interests became more vulnerable to attack.

Domestic Improvements to Homeland Defense

During latter part of the Clinton administration, terrorist attacks on U.S. soil, particularly the World Trade Center bombing, gave impetus to renewed interest in domestic homeland defense capability. But one of the problems the Clinton administration found in establishing an effective domestic homeland defense strategy and policy was the fact that nearly every federal agency or organization has a role in homeland defense. The ANSER Analytic Services organization listed no less than 15 separate federal agencies/organizations (none of which have authority over the other) that have a homeland defense mission.[562] Furthermore, to establish an effective policy and strategy, federal efforts have to be coordinated with state and local governments, as well as private organizations (which control much of the critical power and transportation infrastructure in the U.S.).

Initial efforts to develop a robust domestic WMD preparedness program have not been without problems. Leslie-Anne Levy pointed out that the expansion of federal bureaucracy to deal with the domestic preparedness and response mission resulted in

---

[561] Notes from prepared statements of Joint Chiefs of Staff, unclassified testimony before Senate Armed Service Committee, 26 September 2000.

inevitable duplication of effort and decreased overall effectiveness of programs.[563] For example, though local first responders will often be the first to respond to an attack, they often don't have the money for required equipment and find it difficult to deal with multiple federal agencies.[564] Furthermore, new federal teams trained to respond to WMD incidents have responsibilities that overlap with local first responders.[565] Clearly, the U.S. government is still working on establishing clear missions, responsibilities, and lines of authority in domestic response.

Efforts to revamp and improve domestic WMD attack response capabilities started with Presidential Decision Directive 39 (PDD 39), which broadly defined steps needed to protect government buildings, critical infrastructure, and transportation within the United States from terrorist attacks. But PDD 39 created a problem by insisting that the Department of Justice, through the FBI, became the lead for initial crisis response, and the Federal Emergency Management Agency (FEMA) became the lead in the consequence management phase. The artificial separation of the crisis and consequence management phases of an attack developed a false impression that crisis and consequence management could be neatly separated and compartmentalized.[566] In essence, changing lead agencies in the aftermath of an attack will only confuse and hamper response efforts.

---

[562] ANSER Analytic Services, Inc., *Homeland Defense Federal Organization Agency and Organization Profiles*, (no date), n.p.; on-line, Internet, 30 March 2001, available from http://www.homelanddefense.org/fedorg.cfm.
[563] Leslie-Anne Levy, "Chapter 4: Federal Programs: Disconnected in More Ways Than One," in *Ataxia: The Chemical and Biological Terrorism Threat and the US Response*, Report No. 35 (Washington, D.C.: The Henry L. Stimson Center, October 2000), 113; on-line, Internet, 30 March 2001, available from http://www.stimson.org/pubs/cwc/atxchapter4.pdf.
[564] Loren Thompson, *Homeland Defense: A Confusing Start*, 7 September 1999, 1-2; on-line, Internet, 30 March 2001, available from http://www.defensedaily.com/reports/homeland.htm.
[565] Levy, 113.
[566] Levy, 119.

Additionally, President Clinton issued Presidential Decision Directive 62 (PDD-62), which established the National Coordinator for Security, Infrastructure Protection, and Counter-Terrorism to oversee development of policies and programs to prepare and respond to domestic attacks, including WMD.[567] This document reinforced PDD 39's artificial division of crisis and consequence management. Additionally, it required creation of rapid response teams to assist local responders in the aftermath of a WMD terrorist incident, but PDD 62 did not identify which federal agency should be responsible for these teams. Hence, every agency involved in homeland defense could develop rapid response teams. And the figurehead of National Coordinator for Security, Infrastructure Protection, and Counter-Terrorism had no real authority to execute responsibilities because this Coordinator had no budget authority, and thus had no means of forcing agencies to comply.[568]

Additionally, President Clinton established the President's Commission on Critical Infrastructure Protection (PCCIP). He charged the PCCIP with assessing the threats to and vulnerabilities of our critical infrastructures—defined as "systems whose incapacity or destruction would have a debilitating impact on the defense or economic security of the nation."[569] These critical systems include information and communications, electrical power systems, gas and oil production, storage and transportation; banking and finance;

---

[567] Critical Infrastructure Assurance Office, *Summary of Presidential Decision Directives 62 and 63,* (22 May 1998) 1; on-line, Internet, 9 March 2001, available from http://www.ciao.gov/CIAO_Document_Library/PDD6263_Summary.html.
[568] Levy, 120.
[569] The President's Commission on Critical Infrastructure Protection, *Fact Sheet: President's Commission on Critical Infrastructure Protection* (1997), 1; on-line, Internet, 25 July 2000, available from http://www.info-sec.com/pccip/pccip2/backgrd.html.

transportation; water supply; emergency services; and government services.[570] He also charged the PCCIP with assessing legal and policy issues concerning protection of critical infrastructures, recommending comprehensive national policy and implementation strategy for protecting critical infrastructures, and proposing statutory or regulatory changes to effect recommendations.[571] The PCCIP made several important recommendations for infrastructure assurance and protection, resulting in President Clinton's 1998 Presidential Decision Directive 63 (PDD-63) which outlined the President's policy for Critical Infrastructure Protection.[572] PDD-63 directed development of a national plan for infrastructure protection with initial operating capability by 2000, and by 2003, establishment of a secure, interconnected information systems infrastructure.[573]

The concern over security and support for domestic response to WMD attacks also resulted in an increased role for the Department of Defense. To train soldiers as first responders, the Army established the Domestic Preparedness Program at Soldier and Biological Chemical Command.[574] Additionally, the National Guard and U.S. Joint Forces Command (USJFCOM) developed very important roles in domestic support to WMD attack response efforts.

---

[570] The President's Commission on Critical Infrastructure Protection, *Our Nation's Critical Infrastructures: Some Working Definitions*, (1997) 1-2; on-line, Internet, 25 July 2000, available from http://www.info-sec.com/pccip/pccip2/glossary.html.
[571] President, Executive Order 13010, *President's Commission on Critical Infrastructure Protection* (15 July 1996) 1,3; on-line, Internet, 25 July 2000, available from http://www.info-sec.com/pccip/pccip2/eo13010.html.
[572] President, Presidential Decision Directive 63 (PDD-63), *White Paper: The Clinton Administration's Policy on Critical Infrastructure Protection*, (22 May 1998), 2; on-line, Internet, 25 July 2000, available from http://www.fas.org/irp/offdocs/paper598.htm.
[573] The White House, *Fact Sheet: Protecting America's Critical Infrastructures: PDD 63*, (22 May 1998) 1; on-line, Internet, 25 July 2000, available from http://www.fas.org/irp/offdocs/pdd-63.htm.
[574] Levy, 121.

USJFCOM will now provide forces and capability in support of civil authorities to "manage the consequences of chemical, biological, radiological, nuclear, and enhanced high explosives incidents in the United States."[575] In 1999, USJFCOM established Joint Task Force-Civil Support to integrate all services' domestic terrorism response capabilities.[576]

The National Guard already has a traditional role of supporting state disaster response efforts. The Guard's homeland defense role is an extension of its traditional missions of providing emergency engineering support, security, power, heat, water, transportation, food and shelter. This role now includes "deterring and, when required, defending against strategic attack, supporting civil authorities for crisis management in the event of national response to weapons of mass destruction and ensuring the availability, integrity, survivability, and adequacy of critical national assets."[577]

In 1999, the National Guard began a new effort to respond to WMD terrorism with the establishment of 10 Rapid Assessment and Initial Detection (RAID) teams to help local and state officials respond to terrorist attack. RAID teams were to be comprised of 22 personnel capable of deploying to respond to a terrorist attack within 4 hours. Their responsibilities included identification of biological and chemical agents, tracking dispersal of such agents, and expediting federal and state response to a WMD attack. In 2000, Congress mandated an additional 17 teams for 16 more states. But the initial operating capability of the first RAID teams did not meet the original timeline, and GAO

---

[575] United States Joint Forces Command, *USJFCOM Command Mission*, no date; on-line, Internet, 30 March 2001, available from http://137.246.33.101/cmdmission2.htm.
[576] Levy, 134.
[577] United States Army National Guard, *The Guard Today – Current Initiatives*, no date; on-line, Internet, 30 March 2001, available from http://www.arng.ngb.army.mil/Operations/statements/ps/2001/The%20Guard%20Today.htm.

reports criticized the team's capabilities as redundant and not necessarily helpful if the team arrived 4 hours after an incident. (Note: RAID Teams were redesignated Weapons of Mass Destruction Civil Support Teams in 2000).[578]

The enhanced role of the Department of Defense in domestic homeland defense activities has also become contentious. Civil liberties groups are concerned about *Posse Comitatus* implications of an internal security role for the U.S. military in event of a domestic WMD attack.[579]

Other federal agencies have made significant improvements to their domestic response capabilities, as well. The Department of Health and Human Services (HHS) made some great strides in coordinating medical response to WMD attacks. The HHS plan divided responsibilities for critical aspects of medical response among appropriate agencies (for example, the Centers of Disease Control became responsible for biological agent identification and epidemiological investigation) and gave priority to building up local response capabilities—an important policy move which could help immensely in the first hours after a WMD attack. One important outcome of this locally oriented policy was the establishment of Metropolitan Medical Response System teams to provide immediate medical response after a WMD attack. By 2000, 72 teams existed under this plan. Additionally, HHS established four National Medical Response Teams to help respond to WMD attacks—one special capability these teams carry is enough medicines to treat up to 5000 WMD attack victims.[580] But inspection of medical supplies for these teams uncovered problems in inventory management of basic supplies (surgical gloves)

---

[578] Levy, 140-142.
[579] Jim Landers, "U.S. Quietly Upgrading Homeland Defense Plan," *Dallas Morning News*, 9 February 1999, 2; on-line, Internet, 30 March 2001, available from http://www.devvy.com/homeland/html.
[580] Levy, 124-126.

and maintenance of medicines—a problem that does not bode well for actual response capability in event of WMD attack.[581]

This proliferation of federal bureaucracy for domestic response in some instances created more problems than it solved. The plethora of agencies involved created difficulty in coordinating efforts, created redundancy and waste of funds, and ensured that local first responders have not received as much funding and attention to their needs as some of their federal counterparts.[582] The 4-hour response time for military response teams is probably unrealistic, since dedicated aircraft to this mission are not necessarily available. (For example, the first federal responders to the Oklahoma City bombing did not arrive until 15 hours after the blast).[583] Realistically, local first response teams will be on their own for several hours after the initial attack, and they should be adequately manned and equipped for that contingency.

The federal government's new domestic response policies and programs were tested in the summer of 2000 in an exercise called TOPOFF (an abbreviation for Top Officials) in Denver, Colorado. The exercise tested local, state, and federal response to a simulated bioterrorist attack (release of Yersinia pestis, causative agent of plague) in Denver, Colorado. The results indicated that local resources were soon overwhelmed by the magnitude of the spreading infection. Hospitals did not have enough room to isolate infected patients. Since not enough plague prophylaxis existed to protect the entire population, ad hoc decisions had to be made as to whom would get the prophylaxis. Quarantining the entire 2 million population of Denver was not successful and could not

---

[581] Levy, 126-128.
[582] Levy, 154-157.
[583] Levy, 158.

be enforced.[584] As Dr. Jeff Koplan, Director of the Center for Disease Control, pointed out, "The TOPOFF exercise…illuminated the need for (1) clear quarantine criteria and protocols, (2) clear protocols for local distribution of pharmaceutical stockpiles, (3) increased hospital capacity, and (4) strategies for long-term control of an epidemic."[585]

Clearly, America is not ready for a WMD attack on our soil. The Clinton administration attempted to address the issues, but lack of clear lines of authority and responsibility, limited funding, and the magnitude of the problem of preparing for such an eventuality will continue to challenge administrations to come. If such an attack does occur within the next few years, the initial federal response will most probably be uncoordinated and slow, and local officials will be on their own during the first hours of the crisis. If TOPOFF is any indication of the problems we face in the aftermath of a biological attack, then America could very well face an unstoppable epidemic.

Analysis of American Homeland Defense: The Clinton Era

The Clinton era found the U.S. in a drastically changed threat environment. Gone was the threat of conventional or nuclear war with the Soviet Union and Warsaw Pact—they no longer existed. But threats to the U.S. homeland had not subsided. In fact, the U.S. began to face more diverse and dangerous threats than ever before in history—and the Clinton administration's policies did little to ameliorate them.

    a.   Thanks to the efforts of North Korea, Russia, and China, several nations in the Middle East and Asia now possess ballistic missiles that threaten U.S.

---

[584] Richard E. Hoffman and Jane E. Norton, "Lessons Learned from a Full-Scale Bioterrorism Exercise," *Emerging Infectious Diseases* 6, no. 6 (November-December 2000): n.p.; on-line, Internet, 30 March 2001, available from http://www.cdc.gov/ncidod/eid/vol6no6/hoffman.htm.

[585] Jeff Koplan, M.D., "CDC's Strategic Plan for Bioterrorism," *Biodefense Quarterly* 2, no.3 (December 2000-January 2001): n.p.; on-line, Internet, 30 March 2001, available from http://www.hopkins-biodefense.org/pages/news/quarter.html.

forces overseas and our allies. Some, such as Iraq and Iran, are developing indigenous ballistic missile capability. Some of these states, as well as terrorist groups, either already have or are actively seeking WMD technology and capability. Instead of giving priority to developing the capability to defend the U.S. against ballistic missiles, the Clinton administration cut off funding for space-based missile defense technology in 1993, insisted on abiding by the outdated ABM Treaty, and then deferred a decision on building a missile defense to the next administration. Furthermore, the Clinton administration refused to take effective action against countries that contributed to ballistic missile proliferation (most notably, China). Clinton also failed to develop and execute an effective policy to stop Iraq's WMD program. Hence, the U.S. is more vulnerable than ever before to ballistic missile and WMD attacks; if a robust TMD and NMD capability are not developed and deployed quickly, the U.S. will be vulnerable to ballistic missile attacks from even more countries in the next decade as countries in Asia and the Middle East develop increasingly advanced ballistic missile programs.

b. The U.S. is no safer from terrorist attacks than it was 10 years ago; in fact, terrorist groups' increasing tendency to instigate mass casualty attacks, coupled with acquisition of WMD materials, have made the U.S. even more vulnerable to a devastating terrorist attack within the U.S. The entities posing the greatest terrorist threat to the U.S.—Usama bin Laden, Iraq, and Iran—are still actively engaged in terrorist activity, and the Clinton administration has done little to ameliorate the threats. Clinton's "pin-prick" bombings of Usama bin Laden's terrorist camps and Iraq's intelligence headquarters failed to stop the activities of the persons responsible for terrorist activities against the U.S., and the Clinton administration's alleged "foot dragging" in the Khobar Towers bombing investigation has set a very disturbing precedent in terms of holding state sponsors of terrorism accountable.

c. The Clinton administration consistently failed to match resources to strategic requirements. In addition to numerous "peacekeeping" deployments (such as Somalia, Haiti, Bosnia), the armed forces still had to be prepared to fight two nearly simultaneous MTWs. But budget cuts, lack of force modernization programs, and too many deployments took their toll on U.S. conventional forces capability. Some might perhaps argue that this did not directly affect homeland defense capability in an age where our enemies are more likely to use terrorist attacks or other forms of "asymmetric warfare" against us. But this view is shortsighted. History is replete with lessons of failure to quickly check aggression by aggressive, hegemonic states (Japan's invasion of Manchuria 1931; Germany's reoccupation of the Rhineland in 1936, invasion of Austria in 1938, and annexation of the Sudetenland in 1938).[586] As Cliff Sobel and Loren Thompson pointed out, "The danger lies in the bolder designs an aggressor might pursue if he encounters no resistance."[587] Unless the U.S. is fully prepared to quickly defeat hostile states in any military

---

[586] Dupuy and Dupuy, 1133, 1136, 1145.
[587] Cliff Sobel and Loren Thompson, "The Readiness Trap; The U.S. Military is Failing to Prepare for the Next Big War," *The Heritage Foundation Policy Review*, no. 72 (Spring 1995), 81-83; on-line, Internet, 14 March 2001, available from http://www.policyreview.com/spring95/thompth.html.

confrontation, the U.S. stands to lose access to vital resources, strategic friends and allies, and ability to shape global economic, political, and military events to maintain peace and security in the U.S. Imagine the long-term global consequences of Iranian control of the Persian Gulf, or Saddam Hussein's successful annexation of Kuwait. Unfortunately, the high Clinton-era ops-tempo, combined with military budget and force structure cuts, and failure to fund force modernization, damaged U.S. military readiness and seriously jeopardized our ability to execute two nearly simultaneous MTWs.

d. The Clinton administration started to give serious attention to the problem of preparing and responding to WMD attacks on U.S. soil, but the growth of federal bureaucracies to manage the problem did little to improve response capability, if TOPOFF was any indication of the problems we face. Too many agencies are involved in homeland defense and there is too much redundancy in missions at local, state, and federal levels. As a result, local first responders, who need the most attention, funding, training, and manning, may not receive the support they need for immediate response to a crisis. Projected response times for military response teams are probably not realistic, and are creating unrealistic expectations from local and state officials. And as the TOPOFF exercise proved, we still do not have the required infrastructure or medicines available to effectively respond to a terrorist-induced epidemic.

Chapter 6

Lessons Learned from American Homeland Defense History

Unless history can teach us how to look at the future, the history of war is but a bloody romance.

—Major General J.F.C. Fuller

Although American homeland defense needs continually changed and evolved since the colonial era, three fairly consistent patterns of peacetime homeland defense policy continued up through much of the 20th century: failure to provide the necessary military resources to execute strategic defense requirements, lack of full understanding of the existing threat environment, and failure to develop a viable strategy to protect American vital interests. Generally, only after a crisis emerged did the government hasten to gather the resources and develop a plan to defeat a strategic threat.

The strategy-to-resources imbalance continued throughout succeeding centuries because many civilian and military leaders failed to understand the link between resources and strategy. The problem still exists and is arguably the most pressing issue facing the defense establishment today.

Government parsimony toward expending money for defense needs originated during the colonial era. Colonial (and eventually state) militias jealously maintained control over their militia forces; as a result, readiness of militia forces varied greatly between colonies (and eventually states). Militias were not equipped or trained well enough to handily defeat professional soldiers.

The imbalance between resources and strategic requirements was first apparent in the Revolutionary War. American forces had not the resources nor the experienced leadership required to defeat a powerful enemy with a large, experienced Navy and Army. Not until Washington suffered several defeats in conventional battles with British forces in the Revolutionary War did he realize that his weaker army could only hope to defeat the British using "hit and run" tactics. But even then, French naval assistance was required to win the war.

The War of 1812 again illustrated the American tendency for strategic objectives to outstrip resources. Although the objective of the war was to force the British to respect American naval rights, we didn't have the force structure to achieve our objective, and the British blockaded our harbors. On land, our strategy was no better—the failed invasion of Canada was an ill-conceived notion that wasted resources on an unachievable objective. Again, the French had to come to our rescue to prevent defeat.

The same problem continued throughout the 19th century. After the Mexican War, the U.S. Army did not have the resources to effectively defend the new frontier. During the Civil War, both sides wasted men and material early on due to failure to develop a comprehensive, overarching strategy supportable with available resources. Lack of both military and civilian leadership with an understanding of the connection between strategy and resources plagued both sides—until the Union listened to Winfield Scott and Ulysses S. Grant. They understood the concept of a grand, overarching strategy using all the instruments of national power (economic, political, diplomatic, military, informational) to squeeze the Confederacy into defeat.

The Confederacy, on the other hand, did not develop such a comprehensive design to win the war. The Confederacy lost in part because of Lee's failure to realize that he did not have the resources to develop a workable strategic offensive plan—his only chance was to maintain a defensive strategy to wear down the Union's will to continue the war. His strategic mistake of switching to an offensive strategy doomed the Confederacy at Gettysburg.

Even during the Expansionist Era, when American acquisition of new territories mandated adequate means to defend them, Congress failed to provide the necessary resources. The botched mobilization for the Spanish-American War was a result of continued failure to provide adequate support for our armed forces. The support of Secretaries of the Navy Herbert and Tracy saved the American war effort—if not for their successful push to construct more battleships just prior to the outbreak of the war, the American war effort might have been a failure.

Although the Expansionist Era saw the first glimmerings of American strategic defense planning—with development of the color-coded war plans for defense of U.S. possessions and continental territory—the plans were completely unrealistic, given the limitations on American naval force structure and logistics requirements.

President Wilson's isolationist attitude ensured little attention to military planning or resources in the years leading up to World War I—even as it was clear that America was headed for war. As a result, our military forces were not ready for deployment in World War I. However, despite his desire to keep America out of the war, Wilson did realize that economic and industrial mobilization capacity, as well as authority to use the state militias for national defense purposes, were essential for war preparedness—hence the

National Defense Act of 1916—a key piece of legislation which eased the transition from peacetime to crisis mobilization for later Presidents.

Cuts in defense funding after World War I brought about the same problems in mobilizing for World War II and nearly cost us the Battle of the Atlantic. Additionally, U.S. fixation on the German threat resulted in failure to understand the growing nature of the Japanese threat, culminating in the disaster at Pearl Harbor. World War II did, however, by necessity, give impetus to development of sound joint and combined strategic planning that resulted in victory in both theaters of war.

World War II, the beginning of the Atomic Age, and the beginning of the Cold War forever changed the concept of American homeland defense. Atomic weapons negated any protection America's geographic isolation had previously afforded. With the Soviet Union's nuclear weapons, large standing military force bordering American allies in Europe, vast territory and personnel resources, and intent to spread its influence around the globe at the expense of American vital interests, Soviet actions outside American borders now directly impacted American vital interests and homeland defense capability.

Like it or not, America could never again think of homeland defense in terms of just protecting its territory and population. Homeland defense became much more complex, relying on all instruments of U.S. national power to protect our allies, our access to global resources, our economic strength, control vital lines of communication, and contain the Soviet Union's hegemonic ambitions in nearly every continent.

Unfortunately, successive post-World War II governments continued the previous pattern of holding defense hostage to artificially imposed budget constraints, instead of developing a defense structure appropriate to the threat environment. Furthermore, defense strategy often was not appropriate to the threat environment. The Truman and Eisenhower eras exemplified these problems.

Truman's severe defense cutbacks ensured the U.S. was not ready to prosecute the Korean War. Furthermore, Eisenhower's New Look and New New Look strategies sacrificed flexible, broad range capability on the mistaken belief that atomic weapons were an adequate deterrent to communist threats to U.S. vital interests.

Although Kennedy and McNamara restored some balance to military force structure to enable U.S. forces to fight across the conflict spectrum, neither developed a coherent, effective policy for use of U.S. forces to protect our vital interests. This indirectly brought about the greatest crisis of the Cold War and the most direct threat to the U.S. homeland since World War II—the Cuban Missile Crisis.

But the worst idea regarding homeland defense strategy in the Cold War came from McNamara's intellectually bankrupt idea of "mutual assured destruction" (MAD) as a deterrent to nuclear war. Despite clear evidence that the Soviet Union never accepted this idea and developed both active and passive defenses against nuclear attack, McNamara insisted on using MAD to determine homeland defense strategy and military forces composition; the most far-reaching effect prevented the U.S. from building an anti-ballistic missile capability, even in the face of clear evidence that the Soviets were actively developing such a capability. The ABM Treaty was MAD's ultimate conclusion; successive administrations used the ABM Treaty as a means to prevent development of a viable ballistic missile defense, despite the fact that the Russians violated the treaty by developing an ABM capability.

U.S. nuclear defense capability was not the only program to suffer during the Cold War. Successive Presidents during the Vietnam and post-Vietnam era gave priority to domestic problems, drastically cutting conventional defense forces after the U.S. withdrew from Vietnam, resulting in the now famous "hollow force" of the 1970's.

One improvement in homeland defense strategy that started during the Cold War, however, was the use of security assistance programs to help struggling governments fight internal and external communist threats. The more successful programs, such as the Marshall Plan and Truman's assistance to Greece and Turkey, were very short in duration and had clear, achievable goals. But other assistance programs actually eroded homeland defense capability by expending resources in countries of questionable national interest, forcing the U.S. into open-ended commitments that drained resources from more important programs. Kennedy's decisions to expand U.S. involvement in Vietnam and the Alliance for Progress were two egregious examples.

Furthermore, failure to predict possible unintended consequences of open-ended security assistance programs damaged U.S. credibility and ended up increasing the threat to U.S. homeland security. The escalation of U.S. involvement in the Vietnam War and resultant damage to U.S. defense capability, and the Reagan administration's assistance to Afghan and Nicaraguan rebels both illustrated the dangers of unintended consequences regarding security assistance.

Although security assistance programs became a key element of homeland defense strategy, the Cold War proved these programs must be part of a very clear, overarching strategy for strengthening only specific overseas alliances; furthermore, these programs should be of limited duration and have a clear, achievable goal. But most important, the potential unintended geo-political consequences of security assistance must be analyzed before making any such commitments.

Despite his problems with questionable security assistance programs, President Reagan's was one of the few administrations that developed a clear, specific goal for his national security strategy and secured the resources necessary to execute his strategy—namely, defeat of the Soviet Union. His defense buildup literally restored U.S. homeland defense capability, forced the Soviet Union to agree to a zero-option INF treaty, reversed Soviet geo-political gains, and developed a large, well-equipped, and proficient U.S. armed forces.

Thanks to the Reagan defense buildup, the U.S. led a coalition of nations to fast, decisive defeat of one of the world's largest armies in the 1991 Gulf War. After the Gulf War and the dissolution of the Soviet Union, President Bush, General Colin Powell, and Secretary of Defense Cheney began a reduction of the armed forces to a level more appropriate to the changed threat environment.

But the Clinton administration continued cutting U.S. armed forces even more than Bush had envisioned—while ratcheting up deployment commitments 300 percent and cutting modernization programs. As a result, the U.S. armed forces are now undermanned for the current 2 MTW strategy and are forced to use aging equipment at a time when America's most immediate threats—Usama bin Laden, Iraq, Iran, and North Korea—are continuing to build their capabilities.

Furthermore, in the past 10 years these same countries and bin Laden's group ensured terrorism came to the forefront of threats to the U.S. homeland and our vital interest. The threats posed by Usama bin Laden, Iran, and Iraq illustrate America's continuing failure to develop an adequate homeland defense strategy against terrorism, and to fund the resources necessary to fight this threat. This problem did not develop overnight. America struggled to develop an adequate terrorism policy since the Carter era. The failed Iranian hostage rescue attempt clearly illustrated America's lack of preparedness to fight this new threat. President Reagan's policies were not effective, either, because he lacked consistency in application of policy and failed to develop a long-term, aggressive program to fight terrorism. The one-time attack on Libya did little to damage Libya's capability to support terrorism, and resulted in a retaliatory bombing of a civilian passenger plane. Furthermore, negotiating with terrorists and trading arms for release of American hostages only encouraged more terrorism against Americans.

The Clinton administration did no better. Clinton's failure to persuade (or force) Iran to extradite the Khobar Towers bombers, the one-time attack on Usama bin Laden's terrorist camps in Afghanistan, and the single attack on Iraqi intelligence headquarters illustrated his administration's lack of understanding of the need for unrelenting pressure against terrorists and their state sponsors. Our failure to develop an effective policy or find the resources required to protect the United States and its deployed forces against terrorism have left us vulnerable to further mass-casualty attacks.

Another significant homeland defense policy disaster has been U.S. failure to stop proliferation of ballistic missiles. Thanks to the efforts of Russia, China, and North Korea as exporters of ballistic missile technology, Middle Eastern and Asian nations are quickly developing advanced ballistic missile technology. Thus far, the U.S. has done little to stop proliferation of these technologies—a potentially disastrous oversight in homeland defense policy.

The same problem of failure to develop a coherent policy and strategy for responding to domestic WMD attacks plagued our domestic response capability, as well. The tangle of federal agencies involved in homeland defense preparation and response, lack of adequate resources for local first responders, and the enormous expense of creating the physical infrastructure and medicinal stockpiles needed to successfully treat the victims of a WMD attack has rendered the U.S. highly vulnerable to domestic attack.

Shaping an effective homeland defense policy to defeat threats to the U.S. homeland and its vital interests in this new millennium requires an understanding of the historical mistakes and successes in homeland defense. A review of over three centuries of American homeland defense reveals that the government has often failed to understand the current and emerging threat environment, and lacking that understanding, has failed to develop a clear, effective strategy and provide the resources necessary to defeat these threats. As a result, initial response to a national crisis has often been slow and lacking effectiveness.

Our enemies in this new millennium will not give the U.S. the luxury of months of preparation and mobilization to prepare for a crisis, nor will they fight on our terms. Based on events of the past decade, as well as analysis of emerging threats, our adversaries will avoid confronting the U.S. in a conventional conflict; rather they will use terrorism, ballistic missiles, propaganda, economic and information warfare, and other forms of asymmetric warfare to attack our homeland and vital interests.

For example, China's military leaders see no real "edge" to the battlefield. In war, the battlefield is everywhere—in command, control and communications systems, on the Internet, in the stock market, in the economic center of gravity of the enemy, in the strength or weakness of a country's currency, and in the broadcast media. Future war will blur distinctions between the military and civilians, and new technologies will "[end] the dominance of weapons in war." In essence, anything can become a weapon in war, and traditional distinctions between weapons and non-weapons will be broken. The traditional U.S. advantage in high-tech weaponry will be degraded through these asymmetric attacks.[588]

Hence, our homeland defense strategy must be developed with very clear, specific, achievable goals, based on the current and emerging asymmetric threat environment. Once these goals are determined, the new administration must articulate and fund a military force posture adequate to defeat current and emerging threats. An adequate military force structure must be combined with all the other instruments of national power--political, economic, diplomatic and informational--to provide a truly comprehensive, effective policy for defending the homeland against enemies that will turn our strengths (political, military, and economic alliances with other democracies, strong military force structure, high tech weaponry, interdependent economy and information systems, democratic values and laws) into weaknesses.

---

[588] Qiao Ling and Wang Xiangsui, *Unrestricted Warfare*, U.S. Embassy Beijing summary translation, November 1999, 1-12; on-line, Internet, 13 April 2001, available from http://www.fas.org/nuke/guide/china/doctrine/unresw1.htm.

Chapter 7

Conclusion: American Homeland Defense in the 21st Century

The best way to defeat an enemy is to defeat his strategy. The best way to defeat his strategy is to adopt it.

—Sun Tzu
The Art of War

To develop an effective 21st century homeland defense, the first step is a comprehensive understanding of the current and emerging threat environment. Indirect threats, such as environmental degradation, uncontrolled population growth in countries least able to support it, refugee migration, endemic disease in lesser developed countries, and governmental corruption, coupled with more urgent threats, such as terrorism, ballistic missile and WMD proliferation, hegemonic "rogue" states, and spreading ethnic conflicts which destabilize entire geo-political regions all affect America to a greater or lesser extent due to increasing globalization. Globalization—"the process of accelerating economic, technological, cultural, and political integration"—results in international effects for seemingly local problems.[589]

Globalization has therefore made strictly domestic measures for homeland defense inadequate to address the problem. Homeland defense now requires comprehensive policy and planning among every major branch of the United States government. An effective homeland defense policy and strategy must address domestic vulnerabilities and response capabilities, as well as international measures to protect U.S. vital interests.

Regarding domestic preparedness, the Center for Strategic and International Studies performed a comprehensive study of recent domestic homeland defense efforts and found some critical weaknesses. The most important was lack of an overarching plan for homeland defense because no one had the authority to write it. CSIS recommended that the Vice President be given the job and the authority to develop and coordinate homeland defense efforts by making him (or her) the Chairman of a National Council dedicated to developing and coordinating homeland defense.[590] Another important CSIS recommendation was investment in and enhancement of all-source intelligence collection and analytical capabilities, especially for the purpose of fighting terrorist groups.[591] Some other important CSIS recommendations included networking with scientific and biomedical research communities, tighter coordination among nonproliferation, counter-proliferation, and counter-terrorism communities, investing in chemical, biological, radiological, and nuclear (CBRN) weapons detection and attribution capabilities, improving warning capability, performing an assessment of U.S. intelligence warning capabilities, and required annual net threat assessments for potential CBRN attacks.[592] Implementation of recommendations from the aforementioned CSIS reports should be an integral part of U.S. homeland defense domestic policy and strategy.

However, a comprehensive, overarching strategy for homeland defense should also include using every instrument of U.S. power to shape the international environment to degrade and destroy threats before they ever reach the U.S. homeland. This requires

---

[589] President, *National Security Strategy for a New Century*, (Washington, D.C.: Government Printing Office, December 1999), n.p.; on-line, Internet, 13 February 2001, available from http://ofcn.org/cyber.serv/teledem/pb/2000/jan/msg00037.html.
[590] Joseph J. Collins and Michael Horowitz, *Homeland Defense: A Strategic Approach* (Washington, D.C.: Center for Strategic and International Studies, 2000), 42; on-line, Internet, 9 March 2001, available from http://www.csis.org/homeland/reports/hdstrategicappro.pdf.
[591] Cilluffo et al., 2.

development of clear, specific, international goals to fight current and developing regional and international threats, and providing the military, economic, political, diplomatic, and informational resources to achieve these goals.

Essentially, a comprehensive homeland defense strategy should be fully integrated into the next National Security Strategy—because *the new global threat environment has essentially destroyed any meaningful distinctions between homeland defense and national security.* Threats to U.S. vital interests, people, and territory begin in far-flung areas of the world such as Afghanistan, where international terrorists train and direct operations against the U.S. Therefore, an effective strategy to defeat such threats must also begin with an assessment of international and regional threats, followed by developing both domestic and international homeland defense strategy to defeat these threats.

In his 7 February 2001 testimony to the Senate Select Committee on Intelligence, DCI George Tenet outlined the greatest and most immediate threats to U.S. national security. He stated that Usama bin Laden posed the most immediate and serious threat to U.S. security, based on his 1998 declaration that all Americans were legitimate targets of attack, as well as his campaign of terrorist attacks, and his attacks on Americans since then.[593] In addition to bin Laden, the U.S. and its allies face other terrorist threats as well, including Islamic militancy and Palestinian rejectionist violence, and terrorist acquisition of cyber attack capability and unconventional weapons. Additionally, the U.S. faces ballistic missile threats from Russia, China, North Korea, probably Iran, and possibly Iraq. Tenet was particularly concerned about proliferation of short and medium range

---

[592] Cilluffo et al, 17.
[593] Tenet, "Worldwide Threat 2001: National Security in a Changing World," n.p.

ballistic missiles driven by North Korean, Russian, and Chinese sales[594] (in March 2000 testimony, Tenet stated that MRBM sales, driven particularly by North Korea, were altering the strategic balance of the Middle East).[595] In terms of regional threats, Iraq and Iran (as terrorism sponsors and military threats), and China and Russia (as "would-be" Great Powers) were specifically mentioned as concerns.[596]

If one looks at the plethora of threats facing U.S. vital interests, there are common threads: all are non-democratic, hegemonic "wanna-bes" (including Usama bin Laden, who wants to end U.S. influence in the Middle East to achieve his vision of a pure Islamic culture) who are as yet unable to directly challenge the U.S. militarily, politically, or economically; therefore, they use asymmetric means to attack the U.S. Bin Laden and Iran use terrorism; Iraq uses the classic political ploy of splitting its enemies and weakening their resolve regarding sanctions and WMD inspections; North Korea, China, and Russia fill the Middle East and Asia with ballistic missiles and WMD technology, altering the strategic balance of power against the U.S. and its allies; North Korea fires a ballistic missile over Japan (thereby demonstrating the vulnerability of one of America's key allies); China steals U.S. nuclear secrets; Russia uses political and diplomatic pressure to try to prevent the U.S. from deploying a national missile defense, thereby maintaining an advantage over the U.S in missile defense capability. These are all classic examples of an indirect, asymmetric strategy designed to make the U.S. vulnerable to asymmetric threats—to erode U.S. political power and influence, destroy U.S. capability to project military power and protect itself, split democratic alliances, and gain political

---

[594] Tenet, "Worldwide Threat 2001: National Security in a Changing World," n.p.
[595] Tenet, "The Worldwide Threat in 2000: Global Realities on Our National Security," 1-4.
[596] Tenet, "Worldwide Threat 2001: National Security in a Changing World," n.p.

leverage in strategically critical regions—thereby weakening U.S. ability to defend itself against attack.

A clear, overarching strategy of homeland defense must support a focused objective to defeat these asymmetric threats. The new administration can take a lesson from the success of President Reagan's vision—one which focused the instruments of national power toward one goal—*defeat of the enemy.*

Essentially, our ultimate goal in U.S. homeland defense should be to establish what Frank Gaffney, President of the Center for Security Policy, called a *Pax Democratica*— an international order in which the world's democracies and free market economies actively support fledgling democracies and reform movements in totalitarian countries.[597] In other words, refusing to accept "containment", "engagement," and "stability" as worthy goals, and instead providing universal democratic support for democracies and democratic reform movements. Conversely, this also means using every instrument of national power to defeat the designs of states and entities that threaten the security of the U.S. and its democratic allies.

The Cold War proved only too clearly that the policy of "containment" did not work—it had no clear objective; it resulted in tipping the strategic balance of power against the U.S in the 1970's. Neither will any form of "containment" be adequate to address the threats now facing us. Iraq is a prime example: the U.S has essentially used a containment policy against Iraq for 10 years, and it has failed.

"Engagement and Enlargement" were not adequate to support U.S. homeland security needs in the post-Cold War era, either. The Clinton administration used these concepts as a basis for national security strategy; the result was the increasingly dangerous threat environment we face today. The same states sponsor terrorism and ballistic missile proliferation. Iraq and Iran are growing threats to peace and security in the Middle East. Violence between Israel and the Palestinians has increased and threatens to drag more nations into their conflict. China, Russia, and North Korea continue to arm Asia and the Middle East. China stole U.S. nuclear weapons secrets, continued to build its military arsenal, and threatened Taiwan by launching missiles on the eve of its 1996 Presidential election. North Korea continued its ballistic missile development program and launched a Taepo Dong 1 over Japan. Iraq managed to end the U.N. WMD inspection program and split alliances over the sanctions regime. Usama bin Laden is still free and continuing his terrorism campaign against the U.S., and Iran never turned over the Khobar Towers bombers and has not given up terrorism as an instrument of hegemony. Basically, "engagement and enlargement" became a euphemism for appeasement of dictators.

A homeland defense policy based on *Pax Democratica* would mean more aggressive stance against totalitarian regimes, and it would not be without risk. For example, it would mean actively and publicly supporting Taiwan's economic, military, political security in the face of China's bellicosity. It would mean an end to appeasement and the beginning of a tougher foreign policy against regimes that are a threat to U.S and allied security. In essence, *Pax Democratica* would become a 21st century version of President Reagan's "we win, they lose." Lessons learned from our homeland defense history

---

[597] Frank J. Gaffney, Jr., *Security Policy in the Bush Administration: A Critical Retrospective*, (Washington, D.C.: The Center for Security Policy, October 1992), 3-5; on-line, Internet, 1 February 2001, available from http://www.security-policy.org/papers/studies/bush92.html.

provide a guide for implementing *Pax Democratica* in such a way as to use our strengths against our enemies' weaknesses.

First, accept the fact that we cannot defend against every threat. This means using our resources to defend against the most direct threats to the U.S. homeland while accepting risk in other areas. Develop a comprehensive, overarching strategy that can be implemented using the strengths of our military, political, informational, diplomatic, and economic resources against the weaknesses of our enemies.

Second, the U.S. government has a long history of expecting the military to defend too much with too little, and the current administration can fix this problem by providing the military with enough people and resources to execute whatever strategy civilian leadership deems appropriate. Shortsighted budget limitations, especially in modernization and procurement, will certainly leave the U.S. vulnerable to a growing array of threats over the next decade.

Third, ensure new members of Congress are more literate regarding current national security threats, our strategy for defeating these threats, and the military resources required to defeat current and emerging threats. A start would be to require weeklong national security indoctrination for every new member of Congress. Since members of Congress control the budget, it is important for them to gain an understanding of the threats to U.S. homeland security, the required strategy to defeat these threats, and the defense resources necessary to execute the strategy.

Fourth, develop a clear policy for use of military forces and decide which conflicts and threats are worth using our military resources and which are not. A clear policy for use of military force (such as the Powell Doctrine) will help in developing an appropriate budget for the military and will guide commanders in training for missions. A clear policy on use of force will enhance morale because soldiers will know that when they are deployed, their efforts are essential to national security interests. Conflicts and threats that do not merit use of military resources can and should be handled through other instruments of power.

Fifth, enter no arms control agreements unless they clearly serve one overriding purpose: *enhancement of U.S. homeland security*. Enter no arms control treaties without a means of verification and withdraw from any in which other parties refuse a verification protocol (such as the 1994 Agreed Framework with North Korea). If the other party or parties violate current agreements, give notice of withdrawal. Regarding current agreements, it is imperative for the United States to give the Russians notice of withdrawal from the ABM treaty. This treaty endangers rather than enhances U.S. homeland defense due to the growing diversity of ballistic missile threats; furthermore, the Russians have cheated on the treaty and already possess ABM capability.

Rather than work within the limitations of the legally defunct ABM Treaty, the U.S. must improve sea-based missile defense capability while developing a space-based national missile defense system. The Heritage Foundation's Commission on Missile Defense developed a set of recommendations to begin development of national missile defense capability—first from the sea, then from space.

First and foremost, the Commission recommended removing ABM-related constraints from the Navy's Theater-Wide Missile Defense System. This means getting rid of restraints on use of external sensors to detect missile launches, and linking space-based low-altitude satellite sensors, ground-based radars, and airborne sensors with the

Navy's Aegis missile defense system. Use of external sensors would allow earlier detection of missile launch, thus enhancing effectiveness of a defensive interceptor. Furthermore, the Commission recommended getting rid of restraints on speed of interceptors and permitting a 4.5 kilometer-per-second interceptor (instead of the Clinton administration's artificially imposed slower interceptor speed of 3 kilometers-per-second). A faster interceptor would ensure intercept of an incoming missile much earlier in flight trajectory, thus widening the effective area of defense.[598]

Also, the Commission recommended that the U.S. must convince our allies to support our efforts. The growing ballistic missile threat is not only a danger to the U.S., but our allies as well. Allies could help ensure a viable missile defense system by allowing ground-based radars to be built within their countries, allowing basing support for airborne sensors, and using target tracking information to enhance their own defensive capability.[599]

Finally, the Commission recommended research and development of space-based kinetic energy interceptors and space-based lasers as a follow-on to the Navy's sea-based missile defense systems. A combination of space-based lasers and kinetic energy interceptors would cover the widest possible area coverage for missile defense and provide greater flexibility in defense.[600]

---

[598] The Heritage Foundation Commission on Missile Defense, "Chapter 3: Fundamentals of Global Defense," in *Defending America: A Plan to Meet the Urgent Missile Threat* (Washington, D.C.: The Heritage Foundation, March 1999), 1-13; on-line, Internet, 27 February 2001, available from http://bds.cetin.net.cn:81/cetin2/report/tmd/tmdzl/nmd-US/chapter3.html.
[599] The Heritage Foundation Commission on Missile Defense, "Chapter 4: A Plan for an Affordable and Effective Missile Defense: Recommendations," in *Defending America: A Plan to Meet the Urgent Missile Threat* (Washington, D.C.: The Heritage Foundation, March 1999), 1-18; on-line, Internet, 27 February 2001, available from http://bds.cetin.net.cn:81/cetin2/report/tmd/tmdzl/nmd-US/chapter4.html.
[600] The Heritage Foundation, "Chapter 3: Fundamentals of Global Defense," 5-8.

Until the U.S. can develop and deploy a viable space-based national missile defense, the author of this paper recommends negotiation with our allies to allow deployment of the airborne laser, once it becomes fully operational, in conjunction with Navy area and theater missile defense (TMD) systems, and the Army's Patriot Advanced Capability-3 (PAC-3) and Theater High Altitude Area Defense (THAAD) systems (once fully operational), to selected areas of the globe where the U.S. needs the greatest redundancy and flexibility in missile defense: Northeast Asia, Taiwan, the Mediterranean, and the Persian Gulf. Each of the systems will provide an important capability as part of an overall TMD scheme. The airborne laser system will use a chemical oxygen iodine laser built into a modified Boeing 747-400F. This laser will shoot down multiple enemy ballistic missiles while missiles are still in boost phase over enemy territory, complementing land and naval mid and terminal phase intercept systems by destroying multiple incoming missiles before they reach terminal phase.[601] The Navy's Area Ballistic Missile Defense system will intercept short and medium range ballistic missiles within the atmosphere before land-based missile defense assets arrive in theater (thereby providing protection for sea and aerial ports of debarkation).[602] The Navy's Theater Ballistic Missile Defense will provide exoatmospheric (upper tier) ballistic missile defense against medium and longer range ballistic missiles during ascent, midcourse, and descent phase, providing needed protection before land-based missile defense systems arrive in theater.[603] The PAC-3 will provide lower tier (interception at relatively low

---

[601] Airborne Laser Team, "Airborne Laser Overview," (no date), 1; on-line, Internet, 13 April 2001, available from http://www.airbornelaser.com/special/abl/overview/.

[602] Ballistic Missile Defense Organization, "Navy Area Defense System," (no date), on-line, Internet, 13 April 2001, available from http://www.acq.osd.mil/bmdo/bmdolink/html/navyarea.html.

[603] Ballistic Missile Defense Organization, "Navy Theater Wide Ballistic Missile Defense," BMDO Fact Sheet 202-00-11, (Washington, D.C.: Ballistic Missile Defense Organization, November 2000), 1-2; on-line, Internet, 13 April 2001, available from http://www.acq.osd.mil/bmdo/bmdolink/pdf/aq9903.pdf.

altitudes within the atmosphere)[604] defense against short and medium range ballistic missiles in terminal phase of flight.[605] And the Army's THAAD will provide land-based upper-tier intercept capability against short, medium, and long-range ballistic missiles with intercept either within or outside the atmosphere.[606] Essentially, this would mean development of force structure and doctrine for a joint deployable TMD package comprised of a combination of Navy, Air Force, and Army TMD assets to provide the most flexible, redundant ballistic missile defense possible.

Sixth, review all current security assistance programs, analyze and project unintended consequences, and continue only the ones that truly serve U.S. defense interests, both short- and long-term. Ensure each program has a finite duration and a clearly defined, achievable objective. For example, the U.S. should undertake immediate review of the 1994 Agreed Framework with North Korea and insist on strong verification protocols before any further funds are released for this program.

Seventh, review our alliances and collective security arrangements. If feasible, develop new arrangements with other democratic partners in strategically critical areas as a means of strengthening democratic power and influence at the expense of neighboring totalitarian regimes.

Regarding NATO, our most important alliance, the U.S. must ensure the nations of Central and Eastern Europe know that eventual NATO membership is open to them. Friedbert Pflueger, a member of the German parliament, raised several possibilities for

---

[604] Ballistic Missile Defense Organization, "Theater Missile Defense Programs," (no date); on-line, Internet, 13 April 2001, available from http://www.acq.osd.mil/bmdo/bmdolink/html/tmd.html.
[605] Ballistic Missile Defense Organization, "Patriot Advanced Capability-3," BMDO Fact Sheet 203-00-11, (Washington, D.C.: Ballistic Missile Defense Organization, November 2000), 1-2; on-line, Internet, 13 April 2001, available from http://www.acq.osd.mil/bmdo/bmdolink/pdf/aq9904.pdf.

strengthening and enlargement of the democratic alliances in Europe. Based on their geo-strategic positions, Slovakia and Slovenia should be admitted immediately. Additionally, the U.S. cannot afford to alienate Bulgaria and Romania, or weaken democratic forces within these two key Balkan states, by refusing them any chance of eventual entry into NATO or the European Union. These two countries provide a strategic advantage to NATO through access to the Black Sea and its energy supplies; they must be encouraged to take actions necessary for entry into NATO, or at least the European Union (with eventual eligibility to become part of NATO). Admission of these countries will send a signal of the importance of the Balkan region to Europe's stability and security.[607] Additionally, Russia must not be allowed to prevent integration of the Baltic States (Estonia, Lithuania, and Latvia) into the European Union (or eventually NATO). Admission of the Baltic states will provide a clear signal to Russia that it no longer has a hegemonic hold over these states, and that they are now firmly back within the European family.[608]

In Asia, the U.S. should strengthen political, diplomatic, and military ties with Taiwan and Southeast Asian nations to counteract hegemonic ambitions of China, improve protection of maritime commercial and military assets, and ensure more secure sea lines of communication in the South China Sea, and the littoral areas of Malaysia, Singapore, Indonesia, and Taiwan. The Southeast Asian littoral is of critical importance to world trade and energy resources, and contains one of the most heavily used strategic

---

[606] Ballistic Missile Defense Organization, "Theater High Altitude Area Defense System," BMDO Fact Sheet 204-00-11, (Washington, D.C.: Ballistic Missile Defense Organization, November 2000), 1-2; on-line, Internet, 13 April 2001, available from http://www.acq.osd.mil/bmdo/bmdolink/pdf/aq9905.pdf.
[607] Friedbert Pflueger, "Who's Afraid of Round Two?" *Washington Times*, 19 March 2001, n.p.; on-line, Internet, 19 March 2001, available from http://ebird/dtic.mil/Mar2001/e20011031 9our.htm.
[608] Pflueger, n.p.

waterways of the world—the Strait of Malacca, a narrow waterway between Sumatra and Malaysia. Nearly all commercial sea traffic between Asia, Europe, and the Middle East passes through this narrow strait—including all of the fuel and gas shipments from the Middle East bound for the Far East.[609]

Should China choose to confront the U.S. or other nations in the Southeast Asian area, the Strait of Malacca will be one of the first targets. As Yossef Bodansky, terrorism analyst and a senior consultant for the Department of Defense and Department of State wrote:

> The Strait of Malacca is one of the world's hottest and most crucial strategic choke points. It is considered by experts to be one of the ten most vulnerable objectives which neutralization by hostile forces not only will cause tremendous harm to the well being, perhaps very existence, of the economy of the West, **but** is also very easy to accomplish. Controlling the Strait of Malacca is presently a key strategic objective of the PRC to the point of risking armed conflict with the regional states and even the US.[610]

Access to the Strait of Malacca can be controlled from the Spratly Islands (a group of islands and reefs between Vietnam and the Philippines, north of Malaysia and Indonesia) and Myanmar's (formerly Burma's) coastline on the Bay of Bengal. China is aware of this fact as well, and is supplying weaponry and military infrastructure to Myanmar's government. China also built People's Liberation Army Navy and Marine garrisons on several reefs in the Spratly Islands and conducts naval patrols that interact with the garrisons—actions that have already caused disputes with neighboring nations. However, China's attempts to gain full control of the Spratly Islands and surrounding waters in a

---

[609] Yossef Bodansky, "Beijing's Surge for the Strait of Malacca," (no date), 1; on-line, Internet, 13 April 2001, available from http://www.freeman.org/m_online/bodansky/beijing.htm#N_1_.
[610] Bodansky, 1.

conflict with the U.S. would be greatly limited by their lack of carrier-based air power. But a strong PRC naval capability would not be required to deny access to the Strait of Malacca. Terrorist and pirate attacks, as well as use of Beijing's allies (Iran and Pakistan) to foment Muslim insurrection and attacks on shipping (thereby destabilizing regional governments in Southeast Asia) would achieve the desired effect of crippling economies of nations allied with the U.S. and damaging U.S. credibility and influence in the region.[611]

The entire Southeast Asia littoral area is of vital strategic importance to the economies of many nations, and is a critical sea line of communication for the U.S. It is therefore essential for the U.S. to develop strong political, diplomatic, economic and military ties with neighboring nations to maintain uncontested control of these waterways. Specifically, the U.S. must publicly make a commitment to defense of the Southeast Asian nations against any direct or indirect threat from China or its proxies. Development of stronger military and political ties with India, a growing regional naval power, would be especially important in any potential confrontation in the Southeast Asian littoral.

Eighth, develop and execute a long-term campaign to destroy selected terrorist organizations and punish their state sponsors. The first priority should be destruction of Usama bin Laden and his entire terrorist network. Achieving this goal will require the concerted efforts of the entire defense, intelligence, economic, and foreign policy communities, in concert with some of our trusted allies. Some possible actions to support this goal include a sustained, long term, joint and combined military operation to destroy his organization's international physical infrastructure and personnel, adding Afghanistan

---

[611] Bodansky, 1-4.

to the official list of state sponsors of terrorism (even though the U.S. does not recognize the Taleban as the official government of Afghanistan, it provides safe haven to bin Laden's organization), embargoing trade, withdrawing foreign aid, and cutting diplomatic ties to any nation known to tolerate the presence of members of his organization or hide his money, convincing other countries to support our efforts by freezing his known assets and expelling known members of his organization, and providing selective security and economic incentives to nations which support our goal.

Destroying terrorist organizations goes beyond targeting the groups alone. The states that sponsor these groups must be held accountable for their actions. Through a combination of covert and overt economic, political, military, informational, and diplomatic actions, the U.S. and selected allies can take more forceful measures to discourage the use of terrorism by state sponsors to further their objectives. Iran, Afghanistan, and Iraq are the worst offenders, and should be dealt with accordingly.

Ninth, ensure professional military education includes discussion and planning for dealing with asymmetric warfare. Our next adversary may not be so generous as to give us wide areas of flat terrain to fight on and 6 months to build up our conventional forces. The next major conflict may very well involve a fast-breaking combination of successive high-casualty terrorist attacks on overseas U.S. forces, embassies, and other symbols of American power. Attacks could include cyber attacks on critical command, control, communications, and information networks and keystones of American economic power (creating a stock market crash, hacking into major banks' computers),[612] use of the Internet to create panic and confusion among civilians by spreading rumors of projected biological/chemical attacks, financial disasters, shortages of critical goods), attacks with ballistic missiles carrying WMD on ports, airfields, and areas of U.S. troop concentrations, and a dispersed enemy which hides among civilian population centers. Our professional military education training curriculum (at both service-specific and joint war colleges) should include joint and combined warfare exercises with selected allies to prepare for this type of asymmetric warfare.

Finally, the U.S. government must start giving more funding and resources to our HUMINT community. These are often the most important resources in deterring and fighting asymmetric threats to the U.S. homeland. Many of the recommendations of the Report from the National Commission on Terrorism (often called the "Bremer Commission"), *Countering the Changing Threat of International Terrorism*, could significantly enhance intelligence capability against terrorist groups. Two of the most important and useful recommendations regarding HUMINT included ending restrictions on recruitment of "unsavory sources" and giving higher priority to funding of counter-terrorism efforts by U.S. intelligence and law enforcement agencies.[613]

The ten suggestions listed above do not comprise the sum total of efforts the U.S. should undertake to defend itself. However, they are suggestions based on analysis of both failures and successes in over 300 years of U.S. homeland defense history. Whatever course the new administration takes, time is critical. Unless the U.S. acts now to shape an international environment which enhances the security of the United States, the U.S. will be vulnerable to a 21st century Pearl Harbor.

---

[612] Ling and Xiangsui, 11.
[613] National Commission on Terrorism, n.p.

Glossary

Unless otherwise indicated, all definitions are taken verbatim from the DOD Dictionary of Military Terms, available on-line at http://www.dtic/mil/doctrine/jel/doddict/.

alert. (DOD, NATO) 1. Readiness for action, defense or protection. 2. A warning signal of a real or threatened danger, such as an air attack. 3. The period of time during which troops stand by in response to an alarm. 4. To forewarn; to prepare for action. See also airborne alert. (DOD) 5. A warning received by a unit or a headquarters that forewarns of an impending operational mission. See also air defense warning conditions; ground alert; warning order.

antiterrorism. (DOD) Defensive measures used to reduce the vulnerability of individuals and property to terrorist acts, to include limited response and containment by local military forces. Also called AT. See also antiterrorism awareness; counter-terrorism; proactive measures; terrorism.

arms control. (DOD) A concept that connotes: a. any plan, arrangement, or process, resting upon explicit or implicit international agreement, governing any aspect of the following: the numbers, types, and performance characteristics of weapon systems (including the command and control, logistics support arrangements, and any related intelligence-gathering mechanism); and the numerical strength, organization, equipment, deployment, or employment of the Armed Forces retained by the parties (it encompasses disarmament); and b. on some occasions, those measures taken for the purpose of reducing instability in the military environment.

arms control agreement verification. (DOD) A concept that entails the collection, processing, and reporting of data indicating testing or employment of proscribed weapon systems, including country of origin and location, weapon and payload identification, and event type.

assured destruction. A term used during the Cold War indicating the ability to inflict unacceptable damage on the enemy after the enemy launched a first strike. In numeric terms, it meant capability to destroy 20-25 percent of the Soviet population and about 50 percent of Soviet industry after the Soviet Union had executed a surprise attack on the United States. (Sources: Paret, Peter, ed. *Makers of Modern Strategy from Machiavelli to the Nuclear Age.* Princeton, N.J.: Princeton University Press, 1986; and Weigley, Russell F. *The American Way of War.* Bloomington, Indiana: Indiana University Press, 1973.)

asymmetric engagements. Battles between dissimilar forces. (Source: Joint Pub 1, Joint Warfare of the Armed Forces of the United States. Washington, D.C.: U.S. Government Printing Office, 1995; on-line, available at http://www.adtdl.army.mil/cgi-bin/atdl.dll/jt/1/JP1_ch4.htm#s_34.)

ballistic missile. (DOD, NATO) Any missile that does not rely upon aerodynamic surfaces to produce lift and consequently follows a ballistic trajectory when thrust is terminated. See also aerodynamic missile; guided missile.

centers of gravity. (DOD) Those characteristics, capabilities, or localities from which a military force derives its freedom of action, physical strength, or will to fight. Also called COGs. See also capability.

circular error probable. (DOD) An indicator of the delivery accuracy of a weapon system, used as a factor in determining probable damage to a target. It is the radius of a circle within which half of a missile's projectiles are expected to fall. Also called CEP. See also delivery error; deviation; dispersion error; horizontal error.

civil defense. (DOD) All those activities and measures designed or undertaken to: a. minimize the effects upon the civilian population caused or which would be caused by an enemy attack on the United States; b. deal with the immediate emergency conditions which would be created by any such attack; and c. effectuate emergency repairs to, or the emergency restoration of, vital utilities and facilities destroyed or damaged by any such attack.

consequence management. Includes measures to protect public health and safety, restore essential government services, and provide emergency (relief) to governments, businesses and individuals affected by the consequences of any disaster or emergency situation. (Source: Federal Emergency Management Agency (FEMA) Regional Y2K Workshop Handbook; on-line, available at http://www.fema.gov/y2k/wkshp/handbook.htm.)

contingency. (DOD) An emergency involving military forces caused by natural disasters, terrorists, subversives, or by required military operations. Due to the uncertainty of the situation, contingencies require plans, rapid response, and special procedures to ensure the safety and readiness of personnel, installations, and equipment. See also contingency planning. (Source: Joint Staff Officers Guide AFSC Pub 1—1997; on-line, available at http://www.fas.org/man/dod-101/dod/docs/pub1_97/APPENO.html.)

counterterrorism. (DOD) Offensive measures taken to prevent, deter, and respond to terrorism. Also called CT. See also antiterrorism; combating terrorism; terrorism.

crisis. (DOD) An incident or situation involving a threat to the United States, its territories, citizens, military forces, possessions, or vital interests that develops rapidly and creates a condition of such diplomatic, economic, political, or military importance that commitment of US military forces and resources is contemplated to achieve national objectives. (Source: Joint Staff Officers Guide AFSC Pub 1—1997; on-line, available at http://www.fas.org/man/dod-101/dod/docs/pub1_97/APPENO.html.)

homeland defense. No nationally accepted definition exists. Homeland defense is defined differently by various organizations. Two definitions are provided here:

1. The U.S. Army Training and Doctrine Command defines "homeland defense" as: Homeland defense is protecting our territory, population and critical infrastructure at home by: deterring and defending against foreign and domestic threats; supporting civil authorities for crisis and consequence management; helping to ensure the availability, integrity, survivability, and adequacy of critical national assets.

    (Source: U.S. Army Training and Doctrine Command (TRADOC), *White Paper: Supporting Homeland Defense*, 18 May 1999; on-line, Internet, 1

March 2001, available at http://www.fas.org/spp/starwars/program/homeland/final-white-paper.htm.)

2. The Center for Strategic and International Studies (CSIS) Working Group defines "homeland defense" as:

"...the defense of the United States' territory, critical infrastructure, and population from direct attack by terrorists or foreign enemies operating on our soil..."

(Source: Cilluffo, Frank, et al. *Defending America in the 21st Century*. Washington, D.C.: Center for Strategic and International Studies, 2000. On-line. Internet, 9 March 2001. Available from http://www.csis.org/homeland/reports/defendamer21stexesumm.pdf.)

intercontinental ballistic missile. (DOD) A ballistic missile with a range capability from about 3,000 to 8,000 nautical miles.

intermediate-range ballistic missile. (DOD) A ballistic missile with a range capability from about 1,500 to 3,000 nautical miles.

lines of communications. All the routes--land, water, and air--that connect an operating military force with a base of operations and along which supplies and military forces move. (Joint Pub 4-0) (Source: Joint Staff Officers Guide AFSC Pub 1—1997; on-line, available at http://www.fas.org/man/dod-101/dod/docs/pub1_97/APPENO.html.)

logistics. (DOD) The science of planning and carrying out the movement and maintenance of forces. In its most comprehensive sense, those aspects of military operations which deal with: a. design and development, acquisition, storage, movement, distribution, maintenance, evacuation, and disposition of materiel; b. movement, evacuation, and hospitalization of personnel; c. acquisition or construction, maintenance, operation, and disposition of facilities; and d. acquisition or furnishing of services. (Approved by JMTGM# 061-2846-94) (Source: Joint Staff Officers Guide AFSC Pub 1—1997; on-line, available at http://www.fas.org/man/dod-101/dod/docs/pub1_97/APPENO.html.)

medium-range ballistic missile. (DOD) A ballistic missile with a range capability from about 600 to 1,500 nautical miles.

mutual assured destruction. Term used during the Cold War to describe the fact that both the U.S. and Soviet Union had the capability to inflict unacceptable damage on each other (see "assured destruction"). (Source: _____. "Robert S. McNamara." *SecDef Histories*. On-line. Internet, 8 January 2001. Available from http://www.defenselink.mil/specials/secdef_histories/bios/mcnamara.)

national military strategy. (DOD) The art and science of distributing and applying military power to attain national objectives in peace and war. See also military strategy; national security strategy; strategy; theater strategy.

national objectives. (DOD) The aims, derived from national goals and interests, toward which a national policy or strategy is directed and efforts and resources of the nation are applied. See also military objectives.

national policy. (DOD) A broad course of action or statements of guidance adopted by the government at the national level in pursuit of national objectives.

national security. (DOD) A collective term encompassing both national defense and foreign relations of the United States. Specifically, the condition provided by: a. a military or defense advantage over any foreign nation or group of nations, or b. a favorable foreign relations position, or c. a defense posture capable of successfully resisting hostile or destructive action from within or without, overt or covert. See also security.

national security directive. One of a series of directives that announce Presidential decisions implementing national policy objectives in all areas of national security. All NSDs in this series are individually identified by number and signed by the President. (Source: Joint Staff Officers Guide AFSC Pub 1—1997; on-line, available at http://www.fas.org/man/dod-101/dod/docs/pub1_97/APPENO.html.)

national security interests. (DOD) The foundation for the development of valid national objectives that define US goals or purposes. National security interests include preserving US political identity, framework, and institutions; fostering economic well-being; and bolstering international order supporting the vital interests of the United States and its allies.

national security strategy. (DOD) The art and science of developing, applying, and coordinating the instruments of national power (diplomatic, economic, military, and informational) to achieve objectives that contribute to national security. Also called national strategy or grand strategy. See also military strategy; national military strategy; strategy; theater strategy.

national strategy. (DOD) The art and science of developing and using the political, economic, and psychological powers of a nation, together with its armed forces, during peace and war, to secure national objectives. See also strategy.

nautical mile. (DOD) A measure of distance equal to one minute of arc on the Earth's surface. The United States has adopted the international nautical mile equal to 1,852 meters or 6,076.11549 feet.

nuclear weapon. (DOD, NATO) A complete assembly (i.e., implosion type, gun type, or thermonuclear type), in its intended ultimate configuration which, upon completion of the prescribed arming, fusing, and firing sequence, is capable of producing the intended nuclear reaction and release of energy.

resources. (DOD) The forces, materiel, and other assets or capabilities apportioned or allocated to the commander of a unified or specified command. (JP 1-02) (Source: Joint Staff Officers Guide AFSC Pub 1—1997; on-line, available at http://www.fas.org/man/dod-101/dod/docs/pub1_97/APPENO.html.)

sea-launched ballistic missile. (DOD) A ballistic missile launched from a submarine or surface ship.

security. (DOD) 1. Measures taken by a military unit, an activity or installation to protect itself against all acts designed to, or which may, impair its effectiveness. 2. A condition that results from the establishment and maintenance of protective measures that ensure a state of inviolability from hostile acts or influences. 3. With respect to classified matter, it is the condition that prevents unauthorized persons from having access to official information that is safeguarded in the interests of national security. See also national security.

security assistance. (DOD) Group of programs authorized by the Foreign Assistance Act of 1961, as amended, and the Arms Export Control Act of 1976, as amended, or other related statutes by which the United States provides defense articles, military training, and other defense-related services, by grant, loan, credit, or cash sales in furtherance of national policies and objectives.

short-range ballistic missile. (DOD) A ballistic missile with a range capability up to about 600 nautical miles. Also called SRBM.

special operations. (DOD) Operations conducted by specially organized, trained, and equipped military and paramilitary forces to achieve military, political, economic, or psychological objectives by unconventional military means in hostile, denied, or politically sensitive areas. These operations are conducted during peacetime competition, conflict, and war, independently or in coordination with operations of conventional, nonspecial operations forces. Political-military considerations frequently shape special operations, requiring clandestine, covert, or low visibility techniques and oversight at the national level. Special operations differ from conventional operations in degree of physical and political risk, operational techniques, mode of employment, independence from friendly support, and dependence on detailed operational intelligence and indigenous assets. Also called SO. (JP 1-02) (Source: Joint Staff Officers Guide AFSC Pub 1—1997; on-line, available at http://www.fas.org/man/dod-101/dod/docs/pub1_97/APPENO.html.)

strategic level of war. (DOD) The level of war at which a nation, often as a member of a group of nations, determines national or multinational (alliance or coalition) security objectives and guidance, and develops and uses national resources to accomplish these objectives. Activities at this level establish national and multinational military objectives; sequence initiatives; define limits and assess risks for the use of military and other instruments of national power; develop global plans or theater war plans to achieve these objectives; and provide military forces and other capabilities in accordance with strategic plans. See also operational level of war; tactical level of war.

strategic plan. (DOD) A plan for the overall conduct of a war.

strategic vulnerability. (DOD) The susceptibility of vital elements of national power to being seriously decreased or adversely changed by the application of actions within the capability of another nation to impose. Strategic vulnerability may pertain to political, geographic, economic, scientific, sociological, or military factors.

strategy. (DOD) The art and science of developing and using political, economic, psychological, and military forces as necessary during peace and war, to afford the maximum support to policies, in order to increase the probabilities and favorable consequences of victory and to lessen the chances of defeat. See also military strategy; national strategy.

tactical level of war. (DOD) The level of war at which battles and engagements are planned and executed to accomplish military objectives assigned to tactical units or task forces. Activities at this level focus on the ordered arrangement and maneuver of combat elements in relation to each other and to the enemy to achieve combat objectives. See also operational level of war; strategic level of war.

terrorism. (DOD) The calculated use of unlawful violence or threat of unlawful violence to inculcate fear; intended to coerce or to intimidate governments or societies in the pursuit of goals that are generally political, religious, or ideological. See also antiterrorism; combating terrorism; counterterrorism; terrorist; terrorist groups; terrorist threat conditions.

vulnerability. (DOD) 1. The susceptibility of a nation or military force to any action by any means through which its war potential or combat effectiveness may be reduced or its will to fight diminished. 2. The characteristics of a system which cause it to suffer a definite degradation (incapability to perform the designated mission) as a result of having been subjected to a certain level of effects in an unnatural (manmade) hostile environment. 3. In information operations, a weakness in information system security design, procedures, implementation, or internal controls that could be exploited to gain unauthorized access to information or an information system. See also information; information operations; information system; system.

weapons of mass destruction. (DOD) In arms control usage, weapons that are capable of a high order of destruction and/or of being used in such a manner as to destroy large numbers of people. Can be nuclear, chemical, biological, and radiological weapons, but excludes the means of transporting or propelling the weapon where such means is a separable and divisible part of the weapon. Also called WMD. See also destruction.

Bibliography

_____. *CNN Cold War Episode 24: Conclusions*. On-line. Internet, 4 February 2001. Available from http://www-cgi.cnn.com/SPECIALS/cold.war/episodes/24/script.html.

_____. "Bin Laden Makes a Move." *CBS News*, 13 November 2000. On-line. Internet, 25 February 2001. Available from http://cbsnews.com/now/story/0,1597,206750-412,00.shtml.

_____. "Civil Defense: Evacuous." *The Economist*, 18 November 1978.

_____. "Lebanon History: The Shamun Era 1952-58." On-line. Internet, 10 February 2001. Available from http://www.lebanon.f2s.com/culture/history/independent1.htm.

_____. "Oklahoma Bombing." *Washington Post*, (no date) 1997. On-line, Internet, 23 February 2001. Available from http://washingtonpost.com/wp-srv/national/longterm/oklahoma/oklahoma.htm.

_____. "Robert S. McNamara." *SecDef Histories*. On-line. Internet, 8 January 2001. Available from http://www.defenselink.mil/specials/secdef_histories/bios/mcnamara.

_____. "Text of Fatwa Urging Jihad Against Americans." *Al-Quds al-'Arabi* (in Arabic), 23 February 1998. On-line. Internet, 25 February 2001. Available from http://www.emergency.com/bladen98.htm.

_____. *Reagan and the Soviets: Reagan Doctrine*. On-line. Internet, 12 February 2001. Available from http://www.reagan.dk/newreadoc.htm.

_____. *SecDef Histories—Richard Cheney*. On-line. Internet, 5 February 2001. Available from www.defenselink.mil/specials/secdef_histories/bios/cheney.htm.

Air University. "Chapter 1, Space History." *A War Fighter's Guide to Space: Volume 1*. Maxwell AFB, AL: Air University Press, 1993. On-line. Internet, 12 January 2001. Available from http://www.fas.org/spp/military/docops/usaf/au-18/part01.htm.

Airborne Laser Team. "Airborne Laser Overview." (no date). On-line. Internet, 13 April 2001. Available from http://www.airbornelaser.com/special/abl/overview/.

ANSER Analytic Services, Inc. *Homeland Defense Federal Organization Agency and Organization Profiles*, no date. On-line. Internet, 30 March 2001. Available from http://www.homelanddefense.org/fedorg.cfm.

Austin, Douglas. "Can America Fight Two Wars At Once? That's the Plan, but Experts Doubt It." *Investor's Business Daily*, 28 August 2000.

Ballistic Missile Defense Organization. "Navy Area Defense System." (no date). On-line. Internet, 13 April 2001. Available from http://www.acq.osd.mil/bmdo/bmdolink/html/navyarea.html.

Ballistic Missile Defense Organization. "Patriot Advanced Capability-3." BMDO Fact Sheet 203-00-11. Washington, D.C. Ballistic Missile Defense Organization, November 2000. On-line. Internet, 13 April 2001. Available from http://www.acq.osd.mil/bmdo/bmdolink/pdf/aq9904.pdf.

Ballistic Missile Defense Organization. "Theater High Altitude Area Defense System." BMDO Fact Sheet 204-00-11. Washington, D.C. Ballistic Missile Defense

Organization, November 2000. On-line. Internet, 13 April 2001. Available from http://www.acq.osd.mil/bmdo/bmdolink/pdf/aq9905.pdf.

Ballistic Missile Defense Organization. "Theater Missile Defense Programs." (no date). On-line. Internet, 13 April 2001. Available from http://www.acq.osd.mil/bmdo/bmdolink/html/tmd.html.

Blodgett, Brian. "The Difficulties in the Formation of V Corps for the Spanish-American War." On-line. Internet, 27 October 2000. Available from http://members.tripod.com/Brian_Blodgett/V_Corps_1898.html.

Bodansky, Yossef. "Beijing's Surge for the Strait of Malacca." On-line. Internet, 13 April 2001. Available from http://www.freeman.org/m_online/bodansky/beijing.htm#N_1_.

Bradley, John H. and Jack W. Dice. "The Second World War: Asia and The Pacific." In *The West Point Military History Series*. Edited by Thomas E. Griess. Wayne, NJ: Avery Publishing Group, Inc., 1984.

Brahimi, Rym, Peter Bergen, and David Ensore. "U.S. Finds Link Between bin Laden and Cole Bombing." *Cable News Network*, 7 December 2000. On-line. Internet, 25 February 2001. Available from http://www.cnn.com/2000/US/12/07/cole.suspect/.

Buell, Thomas B. et al. "The Second World War: Europe and the Mediterranean." In *The West Point Military History Series*. Edited by Thomas E. Griess. Wayne, NJ: Avery Publishing Group, Inc., 1984.

Carpenter, Ted Galen. "U.S. Aid to Anti-Communist Rebels: The "Reagan Doctrine" and Its Pitfalls." *Policy Analysis* no. 74 (24 June 1986): page range unknown. On-line. Internet, 12 February 2001. Available from http://www.cato.org/pubs/pas/pa074.html.

Center for Defense Information. *Nuclear Weapons Database: Russian Federation Arsenal: Sea-Based Strategic Weapons: SA-5B Gammon SAM (S-200 Volga)*. 16 November 1998. On-line. Internet, 13 January 2001. Available from http://www.cdi.org/issues/nukef&f/database/rusnukes.html#sa5b.

Center for Defense Information. *Nuclear Weapons Database: Russian Federation Arsenal: Sea-Based Strategic Weapons: SA-10 Grumble SAM (S-300)*. 16 November 1998. On-line. Internet, 13 January 2001. Available from http://www.cdi.org/issues/nukef&f/database/rusnukes.html#sa10.

Center for the Study of Intelligence. *Episode 2, 1963-1965: CIA Judgments on President Johnson's Decision to "Go Big" in Vietnam*. On-line. Internet, 12 February 2001. Available from http://www.cia.gov/csi/books/vietnam/epis2.html.

Cilluffo Frank, et al. *Defending America in the 21st Century*. Washington, D.C.: Center for Strategic and International Studies, 2000. On-line. Internet, 9 March 2001. Available from http://www.csis.org/homeland/reports/defendamer21stexesumm.pdf.

Cincotta, Howard ed. *An Outline of American History*. Washington, D.C.: United States Information Agency, May 1994. On-line. Internet, 22 December 2000. Available from http://usinfo.state.gov/usa/infousa/facts/history/ch7.htm#ambivalent.

Collins, John M. *American and Soviet Military Trends Since the Cuban Missile Crisis*. Washington, D.C.: The Center for Strategic and International Studies, Georgetown University 1978.

Collins, Joseph J. and Michael Horowitz. *Homeland Defense: A Strategic Approach* Washington, D.C.: Center for Strategic and International Studies, 2000. On-line.

Cordesman, Anthony H. "The Military Effectiveness of Desert Fox: A Warning About the Limits of the Revolution in Military Affairs and Joint Vision 2010 (working draft)," 26 December 1998. On-line. Internet, 26 February 2001. Available from http://www.csis.org/stratassessment/reports/effectiveDesertFox.pdf.

Council for a Livable World. *Nuclear Arms Control and the ABM Treaty.* On-line. Internet, 11 January 2001. Available from http://www.clw.org/coalition/nmdbook00abmtreaty.htm.

Critical Infrastructure Assurance Office. *Summary of Presidential Decision Directives 62 and 63* (22 May 1998). On-line. Internet, 9 March 2001. Available from http://www.ciao.gov/CIAO_Document_Library/PDD6263_Summary.html.

Defense Intelligence Agency. "Chapter III, Strategic Defense and Space Programs." In *Soviet Military Power 1985.* On-line. Internet, 8 January 2001. Available from http://www.fas.org/irp/dia/product/smp_85_ch3.htm

Disaster and Emergency Services, Yellowstone County. "History of Emergency Preparedness." On-line. Internet, 24 September 2000. Available from http://www.co.yellowstone.mt.us/des/des_history.asp.

DoD Dictionary of Military Terms. On-line. Internet, 13 March 2001. Available from http://www.dtic.mil/doctrine/jel/doddict/.

DoD News Briefing. "Pentagon Details U.S. Missile Attack on Iraq." 26 June 1993. On-line. Internet, 26 February 2001. Available from http://www.fas.org/man/dod-101/ops/docs/dod_930626.htm.

Donovan, Timothy H., Jr., et al. "The Civil War." In *The West Point Military History Series.* Edited by Thomas E. Griess. Wayne, New Jersey: Avery Publishing Group Inc., 1987.

Dougherty, Chuck. "The Minutemen, The National Guard, and the Private Militia Movement: Will the Real Militia Please Stand Up?" *John Marshall Law Review*, (Summer 1995): page range unknown. On-line. Internet, 20 December 2000. Available from http://www.saf.org/LawReviews/Dougherty1.html.

Dupuy, R. Ernest, and Trevor N. Dupuy. *The Harper Encyclopedia of Military History from 3500 B.C. to the Present.* 4th ed. New York: Harper Collins Publishers, 1993.

Eisenhower, Dwight D. "Eisenhower Doctrine 1957." 5 January 1957. *Public Papers of the Presidents, Dwight D. \cf2 Eisenhower\cf0, 1957.* On-line. Internet, 10 February 2001. Available from http://coursesa.matrix.msu.edu/~hst306/documents/eisen.html.

Fabyanic, Thomas A. "The U.S. Air Force." In *Defense Policy in the Reagan Administration.* Washington, D.C.: National Defense University Press 1988.

Faruqi, Anwar. "Russia Arming An Iran In Disputes With Almost All Its Neighbors." *Associated Press*, 20 March 2001. On-line. Internet, 20 March 2001. Available from http://ebird/dtic.mil/Mr2001/e20010320arming.htm.

Federal Bureau of Investigation, Counter-terrorism Threat Assessment and Warning Unit, National Security Division. *Terrorism in the United States 1996.* On-line. Internet, 23 February 2001. Available from http://www.fbi.gov/library/terror/terroris.pdf.

Federation of American Scientists, *R-36O / SL-X-? FOBS.* On-line. Internet, 12 January 12 2001. Available from http://www.fas.org/nuke/guide/russia/icbm/r-36o.htm.

Federation of American Scientists, *S-300V SA-12A GLADIATOR and SA-12B GIANT HQ-18*. 30 June 2000. On-line. Internet, 13 January 2001. Available from http://www.fas.org/nuke/guide/russia/airdef/s-300v.htm.

Federation of American Scientists, *Vietnam War*. On-line. Internet, 12 February 2001. Available from http://www.fas.org/man/dod-101/ops/vietnam.htm.

Federation of American Scientists. *Afghanistan—Introduction*. On-line. Internet, 17 February 2001. Available from http://www.fas.org/irp/world/afghan/intro.htm.

Federation of American Scientists. *Anti-Ballistic Missile Treaty Chronology*. On-line. Internet, 12 January 2001. Available from http://www.fas.org/nuke/control/abmt/chron.htm.

Federation of American Scientists. *B-1A*. 25 March 1998. On-line. Internet, 26 January 2001. Available from http://www.fas.org/nuke/guide/usa/bomber/b-1a.htm.

Federation of American Scientists. *R-36 / SS-9 SCARP*. On-line. Internet, 10 January 2001. Available from http://www.fas.org/nuke/guide/russia/icbm/r-36.htm.

Federation of American Scientists. *Strategic Arms Reduction Treaty (START I) Chronology*. On-line. Internet, 4 February 2001. Available from http://www.fas.org/nuke/control/start1/chron.htm.

Federation of American Scientists. *UR-100 / SS-11 SEGO*. On-line. Internet, 10 January 2001. Available from http://www.fas.org/nuke/guide/russia/icbm/ur-100k.htm.

Finley, James P. "Buffalo Soldiers at Huachuca: Villa's Raid on Columbus, New Mexico." *Huachuca Illustrated* 1, (1993): n.p. On-line. Internet, 9 November 2000. Available from http://www.ukans.edu/~kansite/ww_one/comment/huachuca/HI1-12.htm.

Flint, Roy K. Peter W. Kozumplik, and Thomas J. Waraksa. "The Arab-Israeli Wars, The Chinese Civil War, and The Korean War." In *The West Point Military History Series*. Edited by Thomas E. Griess. Wayne, NJ: Avery Publishing Group, Inc., 1987.

Foerster, Schuyler. "Arms Control: Redefining the Agenda." In *Defense Policy in the Reagan Administration*. Washington, D.C.: National Defense University Press 1988..

Fox, Jonathan. *United States Foreign Policy in the Twenty-First Century: The Crisis and Renewal of the Republican Empire*. No date. On-line. Internet, 9 February 2001. Available from http://www.spaef.com/JPE_PUB/vln3_fox.html.

Fuller, J. F. C. *A Military History of the Western World, Volume III, From the American Civil War to the End of World War II*. New York: Da Capo Press, 1956.

Gaffney, Frank J. Jr. *Security Policy in the Bush Administration: A Critical Retrospective*. Washington, D.C.: The Center for Security Policy (October 1992). On-line. Internet, 1 February 2001. Available from http://www.security-policy.org/papers/studies/bush92.html.

Garrison, Dee. *Civil Defense Portrayal of Nuclear War*. PBS interview. On-line. Internet, 25 September 2000. Available from http://www.pbs.org/wgbh/amex/bomb/filmmore/refernce/interview/garrison2.html.

Goodpaster, Andrew. *Eisenhower's Civil Defense Program*. PBS interview. On-line. Internet, 6 January 2001. Available from http://www.pbs.org.wgbh/amex/bomb/filmmore/reference/interview/goodpaster02.html.

Goure, Daniel and Jeffrey M. Ranney. *Averting the Defense Train Wreck in the New Millennium*. Washington, D.C.: The CSIS Press, 1999.

Grinter, Lawrence E. *Avoiding the Burden: The Carter Doctrine in Perspective*. On-line. Internet, 12 February 2001. Available from www.airpower.maxwell.af.mil/airchronicles/aureview/1983/jan-feb/grinter.html.

Guilmartin, John F., Jr. "Terrorism: Political Challenge and Military Response." In *Defense Policy in the Reagan Administration*. Edited by William P. Snyder and James Brown. Washington, D.C.: National Defense University Press 1988.

Hansell, Haywood S., Jr. *The Strategic Air War Against Germany and Japan: A Memoir*. Washington, D.C.: Government Printing Office, 1986.

Hitch, Charles J. *Decision-Making for Defense*. Berkeley, California: University of California Press, 1965.

Hoffman, Richard E. and Jane E. Norton. "Lessons Learned from a Full-Scale Bioterrorism Exercise." *Emerging Infectious Diseases* 6, no. 6 (November-December 2000): n.p. On-line. Internet, 30 March 2001. Available from http://www.cdc.gov/ncidod/eid/vol6no6/hoffman.htm.

Holmes Kim R. et al. "Preface." *Defending America : A Plan to Meet the Urgent Missile Threat*. Heritage Foundation Report. Washington, D.C.: Heritage Foundation Commission on Missile Defense, March 1999. On-line. Internet, 27 February 2001. Available from http://www.heritage.org/missile_defense/preface.html.

Holmes, Kim R. et al. "Chapter 2 The ABM Treaty and Intentional Vulnerability." *Defending America : A Plan to Meet the Urgent Missile Threat*. Heritage Foundation Report, Washington, D.C.: Heritage Foundation Commission on Missile Defense, March 1999. On-line. Internet, 27 February 2001. Available from http://www.heritage.org/missile_defense/chapter2.html.

House. *U.S. National Missile Defense Policy and the Anti-Ballistic Missile Treaty: Hearings Before the Committee on Armed* Services. 106th Cong., 1st sess., 13 October 1999. On-line, 13 January 2001. Available from http://commdocs.house.gov/committees/security/has286000.000/has286000_0.HTM.

Ippolito, Dennis S. "Defense Budgets and Spending Control: The Reagan Era and Beyond." In *Defense Policy in the Reagan Administration*. Edited by William P. Snyder and James Brown. Washington, D.C.: National Defense University Press 1988.

Jordan, Amos A., William J. Taylor, Jr., and Lawrence J. Korb. *American National Security: Policy and Process*. 3rd ed. Baltimore, MD: Johns Hopkins University Press, 1989.

Kagan, Robert. "The Clinton Legacy Abroad." *Weekly Standard*, 15 January 2001. On-line. Internet, 16 January 2001. Available from http://dailyread/esup/tues/s20010116legacy.htm.

Keegan, John. *The Mask of Command*. New York: Viking Penguin Inc., 1987.

Kennedy, John F., President. Inaugural Address. January 20, 1961. On-line. Internet, 25 September 2000. Available from http://www.bartleby.com/124/pres56.html.

Kennedy, John F., President. "Special Message to Congress on Urgent National Needs." 25 May 1961. On-line. Internet, 25 September 2001. Available from http://www.cs.umb.jfklibrary/j052561.htm.

Kinnard, Douglas. *The Secretary of Defense*. Lexington, Kentucky: The University Press of Kentucky, 1980.

Koplan, Jeff, M.D. "CDC's Strategic Plan for Bioterrorism." *Biodefense Quarterly* 2, no.3 (December 2000-January 2001): n.p. On-line. Internet, 30 March 2001. Available from http://www.hopkins-biodefense.org/pages/news/quarter.html.

Korb, Laurence J. "The Defense Policy of the United States." In *The Defense Policies of Nations: A Comparative Study*. Edited by Douglas J. Murray and Paul R. Viotti. Baltimore, MD: Johns Hopkins University Press, 1982.

Korb, Lawrence J. "The Defense Policy of the United States." In *The Defense Policies of Nations: A Comparative Study*. Edited by Douglas J. Murray and Paul R. Viotti. Baltimore, MD: Johns Hopkins University Press 1982.

Korb, Lawrence J. *U.S. National Defense Policy in the post-Cold War World*. 14 June 2000. On-line. Internet, 1 February 2001. Available from http://www.foreignrelations.org/public/armstrade/korb_postcoldwar_paper.html.

Krasnoborski, Edward J. and George Giddings. "Atlas for The Second World War: Asia and the Pacific." In *The West Point Military History Series*. Edited by Thomas E. Griess. Wayne, NJ: Avery Publishing Group, Inc., 1984.

Landers, Jim. "U.S. Quietly Upgrading Homeland Defense Plan." *Dallas Morning News*, 9 February 1999. On-line. Internet, 30 March 2001. Available from http://www.devvy.com/homeland/html.

Lee, William T. *Ballistic Missile Defense and Arms Control Follies*. 25 September 1996. On-line. Internet, 9 January 2001. Available from http://www.fas.org/spp/starwars/congress/1996_h/h960927l.htm.

Levy, Leslie-Anne. "Chapter 4: Federal Programs: Disconnected in More Ways Than One." In *Ataxia: The Chemical and Biological Terrorism Threat and the US Response*. Report No. 35. Washington, D.C.: The Henry L. Stimson Center, October 2000. On-line. Internet, 30 March 2001. Available from http://www.stimson.org/pubs/cwc/atxchapter4.pdf.

Library of Congress. "A Survey of the Marshall Plan and its Consequences." *Marshall Plan Exhibit: For European Recovery: The Fiftieth Anniversary of the Marshall Plan*. On-line. Internet, 10 February 2001. Available from http://lcweb.loc.gov/exhibits/marshall/m41.html.

Library of Congress. "The Marshall Plan Countries." *Marshall Plan Exhibit: For European Recovery: the Fiftieth Anniversary of the Marshall Plan*. On-line. Internet, 10 February 2001. Available from http://lcweb.loc.gov/exhibits/marshall/mars5.html.

Ling, Qiao and Wang Xiangsui, *Unrestricted Warfare*, U.S. Embassy Beijing summary translation, November 1999, 11; on-line, Internet, 13 April 2001, available from http://www.fas.org/nuke/guide/china/doctrine/unresw1.htm.

Mandelbaum, Michael. "American Policy: The Luck of the President." *Foreign Affairs* 64, no. 3 (1986): page range unknown.

Martin, David C. and John Walcott. *Best Laid Plans: The Inside Story of American's War Against Terrorism*. New York: Harper and Row, 1988.

McCarthy, Andrew C. "Prosecuting the New York Sheikh." *Middle East Quarterly* (March 1997): page range unknown. On-line. Internet, 25 February 2001. Available from http://www.ict.org.il/articles/articledet.cfm?articleid=95.

McCarthy, Michael J. "Lafayette, We Are Here: The War College Division and American Military Planning for the AEF in World War I." Master's thesis, Marshall University, 1992. On-line. Internet, 25 October 2000. Available from http://mccarthy.marshall.edu/thesis/aef_2.txt.

McEnaney, Laura. *America's Evacuating Cities*. PBS Interview. On-line. Internet, 6 January 2001. Available from http://www.pbs.org/wgbh/amex/bomb/filmmore/reference/interview/mcenaney05.html.

McIntyre, Jamie. "Moderate to Severe Damage Seen at suspected bin Laden Camps." *Cable News Network*, 13 January 1999. On-line. Internet, 25 February 2001. Available from http://www.cnn.com/WORLD/asiapcf/9901/13/afghan.damage.photos/.

McMaster, H.R. "Graduated Pressure: President Johnson and the Joint Chiefs." *Joint Forces Quarterly* (Autumn/Winter 1999-2000): page range unknown. On-line. Internet, 12 February 2001. Available from http://www.dtic.mil/doctrine/jel/jfq_pubs/1723.pdf.

McNamara, Robert S. Quoted by Jack Swift, "Strategic Superiority Through SCI." *Defense and Foreign Affairs*, December 1985.

McNamara, Robert S., Secretary of Defense. Address at Ann Arbor, Michigan, June 1962. On-line. Internet, 15 September 2000. Available from http://www.nuclearfiles/org/docs/1962/620709-mcnamara.html.

Merriam-Webster's Collegiate Dictionary. "Fatwa." On-line. Internet, 25 February 2001. Available from http://www.m-w.com/cgi-bin/dictionary?book=Dictionary&va=fatwa.

Millett, Allan R. and Williamson Murray, ed. *Military Effectiveness Volume II: The InterWar Period*. Boston: Unwin Hyman, Inc., 1988.

National Archives and Records Administration. *Constitution of the United States of America*. On-line. Internet. 17 February 2001. Available from http://www.nara.gov/exhall/charter/consitution/constitution.html.

National Commission on Terrorism. *Countering the Changing Threat of International Terrorism*. 5 June 2000. On-line. Internet, 3 February 2001. Available from http://www.fas.org/irp/threat/commission.html.

National Security Council. *NSC 68: United States Objectives and Programs for National Security*. Washington, D.C.: National Security Council, 14 April, 1950. On-line. Internet, 6 January 2001. Available from http://www.fas.org/irp/offdocs/nsc-hst/nsc-68-cr.htm.

National War College. *NWC Course 5605, Military Strategy and Operations, Topic 12: Homeland Defense*, 24 March 2000. On-line. Internet. 24 March 2000. Available from http://www.ndu.edu/ndu/nwc/5605SYL/Topic12.html.

Newhouse, John. *War and Peace in the Nuclear Age*. New York: Alfred A. Knopf, 1988.

Notes from prepared statements of Joint Chiefs of Staff. Unclassified testimony before Senate Armed Service Committee, 26 September 2000.

Novichkov, Nikolay and Michael Dornheim. "Russian SA-12, SA-10 On World ATBM Market," *Aviation Week and Space Technology* 146, no.9 (3 March 1997): 58.

Office of the Assistant Secretary of Defense (Public Affairs). "Iraq's Chemical and Biological Weapons Capability." Briefing, 14 November 1997. On-line. Internet, 26 February 2001. Available from http://www.defenselink.mil/news/Nov1997/x11171997_x114iraq.html.

Office of the Chief of Military History, United States Army. "Chapter 1, Introduction." In *American Military History*. Washington, D.C.: Office of the Chief of Military History, 1988. On-line. Internet. 30 November 2000. Available from http://www.army.mil/cmh-pg/books/amh/amh-01.htm.

Office of the Chief of Military History, United States Army. "Chapter 2, The Beginnings." In *American Military History*. Washington, D.C.: Office of the Chief of Military History, 1988. On-line. Internet. 30 November 2000. Available from http://www.army.mil/cmh-pg/books/amh/amh-02.htm.

Office of the Chief of Military History, United States Army. "Chapter 3, The American Revolution: First Phase." In *American Military History*. Washington, D.C.: Office of the Chief of Military History, 1988. On-line. Internet. 30 November 2000. Available from http://www.army.mil/cmh-pg/books/amh/amh-03.htm.

Office of the Chief of Military History, United States Army. "Chapter 4, The Winning of Independence, 1777-1783." In *American Military History*. Washington, D.C.: Office of the Chief of Military History, 1988. On-line. Internet. 30 November 2000. Available from http://www.army.mil/cmh-pg/books/amh/amh-04.htm.

Office of the Chief of Military History, United States Army. "Chapter 6, The War of 1812." In *American Military History*. Washington, D.C.: Office of the Chief of Military History, 1988. On-line. Internet. 30 November 2000. Available from http://www.army.mil/cmh-pg/books/amh/amh-06.htm.

Office of the Chief of Military History, United States Army. "Chapter 7, The Thirty Years' Peace." In *American Military History*. Washington, D.C.: Office of the Chief of Military History, 1988. On-line. Internet. 30 November 2000. Available from http://www.army.mil/cmh-pg/books/amh/amh-07.htm.

Office of the Chief of Military History, United States Army. "Chapter 8, The Mexican War and After." In *American Military History*. Washington, D.C.: Office of the Chief of Military History, 1988. On-line. Internet. 30 November 2000. Available from http://www.army.mil/cmh-pg/books/amh/amh-08.htm.

Office of the Chief of Military History, United States Army. "Chapter 9, The Civil War, 1861." In *American Military History*. Washington, D.C.: Office of the Chief of Military History, 1988. On-line. Internet. 30 November 2000. Available from http://www.army.mil/cmh-pg/books/amh/amh-09.htm.

Office of the Chief of Military History, United States Army. "Chapter 10, The Civil War 1862." In *American Military History*. Washington, D.C.: Office of the Chief of Military History, 1988. On-line. Internet. 30 November 2000. Available from http://www.army.mil/cmh-pg/books/amh/amh-10.htm.

Office of the Chief of Military History, United States Army. "Chapter 11, The Civil War 1863." In *American Military History*. Washington, D.C.: Office of the Chief of Military History, 1988. On-line. Internet. 30 November 2000. Available from http://www.army.mil/cmh-pg/books/amh/amh-11.htm.

Office of the Chief of Military History, United States Army. "Chapter 12, The Civil War 1864-1865." In *American Military History*. Washington, D.C.: Office of the Chief of Military History, 1988. On-line. Internet. 30 November 2000. Available from http://www.army.mil/cmh-pg/books/amh/amh-12.htm.

Office of the Chief of Military History, United States Army. "Chapter 13, Darkness and Light: The Interwar Years, 1865-1898." In *American Military History*. Washington, D.C.: Office of the Chief of Military History, 1988. On-line. Internet. 30 November 2000. Available from http://www.army.mil/cmh-pg/books/amh/amh-13.htm.

Office of the Chief of Military History, United States Army. "Chapter 15, Emergence to World Power 1898-1902." In *American Military History*. Washington, D.C.: Office of the Chief of Military History, 1988. On-line. Internet. 30 November 2000. Available from http://www.army.mil/cmh-pg/books/amh/amh-15.htm.

Office of the Chief of Military History, United States Army. "Chapter 17, World War I: The First Three Years." In *American Military History*. Washington, D.C.: Office of the Chief of Military History, 1988. On-line. Internet. 26 October 2000. Available from http://www.army.mil/cmh-pg/books/amh/amh-17.htm.

Office of the Chief of Military History, United States Army. "Chapter 19, Between World Wars." In *American Military History*. Washington, D.C.: Office of the Chief of Military History, 1988. On-line. Internet. 26 October 2000. Available from http://www.army.mil/cmh-pg/books/amh/amh-19.htm.

Office of the Chief of Military History, United States Army. "Chapter 26, The Army and the New Look." In *American Military History*. Washington, D.C.: Office of the Chief of Military History, 1988. On-line. Internet. 5 January 2001. Available from http://www.army.mil/cmh-pg/books/amh/amh-26.htm.

Office of the Chief of Military History, United States Army. "Chapter 24, Peace Becomes Cold War, 1945-1950." In *American Military History*. Washington, D.C.: Office of the Chief of Military History, 1988. On-line. Internet. 30 November 2000. Available from http://www.army.mil/cmh-pg/books/amh/amh-24.htm.

Office of the Chief of Military History, United States Army. "Chapter 27, Global Pressures and the Flexible Response." In *American Military History*. Washington, D.C.: Office of the Chief of Military History, 1988. On-line. Internet. 30 November 2000. Available from http://www.army.mil/cmh-pg/books/amh/amh-27.htm.

Olson, Kyle B. "Aum Shinrikyo: Once and Future Threat?" *Emerging Infectious Diseases*, 5, no. 4 (July-August 2000): page range unknown. On-line. Internet, 23 February 2001. Available from www.cdc.gov/ncidod/EID/vol5no4/olson.htm.

Paret, Peter, ed. *Makers of Modern Strategy from Machiavelli to the Nuclear Age*. Princeton, N.J.: Princeton University Press, 1986.

Pflueger, Friedbert. "Who's Afraid of Round Two?" *Washington Times*, 19 March 2001. On-line. Internet, 19 March 2001. Available from http://ebird/dtic.mil/Mar2001/e20011030319our.htm.

Possony, Stefan T., Jerry E. Pournelle, and Colonel Francis X. Kane. *The Strategy of Technology*. Electronic edition, 1997. On-line. Internet, 15 September 2000. Available from http://www.jerrypournelle.com/sot/sot_6.htm.

President. *A National Security Strategy of Engagement and Enlargement 1996.* Washington, D.C.: Government Printing Office, February 1996. On-line. Internet, 21 February 2001. Available from http://www.fas.org/spp/military/docops/national/1996stra.htm.

President. Executive Order 13010. *President's Commission on Critical Infrastructure Protection* (15 July 1996). On-line. Internet, 25 July 2000. Available from http://www.info-sec.com/pccip/pccip2/eo13010.html.

President. *National Security Decision Directive Number 32, U.S. National Security Strategy.* 20 May 1982. On-line. Internet, 17 January 2001. Available from http://www.fas.org/irp/offdocs/nsdd/23-1618t.gif.

President. *National Security Strategy for a New Century.* Washington, D.C.: Government Printing Office, December 1999. On-line. Internet, 13 February 2001. Available from http://ofcn.org/cyber.serv/teledem/pb/2000/jan/msg00037.html.

President. *National Security Strategy of the United States.* Washington, D.C.: Government Printing Office, August 1991. On-line. Internet, 1 February 2001. Available from http://www.fas.org/man/docs/918015-nss.htm.

President. Presidential Decision Directive 63 (PDD-63). *White Paper: The Clinton Administration's Policy on Critical Infrastructure Protection* (22 May 1998). On-line. Internet, 25 July 2000. Available from http://www.fas.org/irp/offdocs/paper598.htm.

Reagan, Ronald, President. Quoted in *Reagan and the Soviets: Reagan Doctrine.* On-line. Internet, 12 February 2001. Available from http://www.reagan.dk/newreadoc.htm.

Reisman, W. Michael. "The Raid on Baghdad: Some Reflections on its Lawfulness and Implications." *American Journal of International Law*, 5, no. 1 (no date): page range unknown. On-line. Internet, 26 February 2001. Available from http://www.ejil.org/journal/Vol5/No1/art11.html.

Robinson, Julian Perry and Jozef Goldblat. *Chemical Warfare in the Iraq-Iran War.* Stockholm International Peace Research Institute (SIPRI) Fact Sheet (Stockholm, Sweden: SIPRI, 1984. On-line. Internet, 13 March 2001. Available from http://projects.sipri.se/cbw/research/factsheet-1984.html.

Rumsfeld, Donald H. et al. *Executive Summary of the Report of the Commission to Assess the Ballistic Missile Threat to the United States.* 104th Cong., 15 July 1998. On-line. Internet, 26 February 2001. Available from http://www.fas.org/irp/threat/bm-threat.htm.

Sale, Richard. "Terrorists Targeted Disneyland, Space Needle." *United Press International*, 20 February 2001.

Scarborough, Rowan. "Readiness of Armed Forces is Not Improving; Clinton Action on Pentagon Cuts Seen as Cause of Problem." *The Washington Times*, 28 August 2000.

Schoenherr, Steve, University of San Diego History Department. *Ford-Carter Era: 1974-1980.* On-line. Internet, 15 January 2001. Available from http://history.acusd.edu/gen/20th/carter.html.

Scott, James M. "Interbranch Rivalry and the Reagan Doctrine in Nicaragua." *Political Science Quarterly*, Summer 1997, 1-10. On-line. Internet, 16 February 2001. Available from www.britannica.com/bcom/magazine/article/0,5744,237514,00.html.

Senate. Resolution 408, 100th Cong., 2nd sess., 24 June 1988. On-line. Internet, 26 February 2001. Available from http://www.senate.gov/~rpc/rva/1002/1002201.htm.

Skidmore, Thomas E. and Peter H. Smith. "The Rise of U.S. Influence." In *Modern Latin America*. 2nd ed. Oxford University Press, 1989. On-line. Internet, 10 February 2001. Available from http://www.mty.itesm.mx/dch/deptos/ri/ri-802/lecturas/lecvmx024.html.

Snyder, William P. and James Brown, ed. "Introduction." In *Defense Policy in the Reagan Administration*. Washington, D.C.: National Defense University Press 1988.

Sobel, Cliff and Loren Thompson. "The Readiness Trap; The U.S. Military is Failing to Prepare for the Next Big War." *The Heritage Foundation Policy Review*, no. 72 (Spring 1995): n.p. On-line. Internet, 14 March 2001. Available from http://www.policyreview.com/spring95/thompth.html.

Spector, Ronald. "The Military Effectiveness of the US Armed Forces, 1919-39." In *Military Effectiveness Volume II: The Interwar Period*. Edited by Allan R. Millett and Williamson Murray. Boston: Unwin Hyman, 1988.

Spence, Floyd, Chairman, House National Security Committee. "Unveiling the Ballistic Missile Threat: The Ramifications of the Rumsfeld Report." *National Security Report*, 2, issue 4 (August 1998): no pagination. On-line. Internet, 27 February 2001. Available from http://www.house.gov/hasc/Publications/105thCongress/NSRs/nsr2-4rumsfeldreport.pdf.

Spring, Baker. "Clinton's Failed Missile Defense Policy: A Legacy of Missed Opportunities." *The Heritage Foundation Backgrounder* no. 1396 (21 September 2000): n.p. On-line. Internet, 21 February 2001. Available from http://www.heritage.org/library/backgrounder/bg1396.html.

Statement of John Foster Dulles, Secretary of State. In "The Evolution of Foreign Policy." *Department of State Bulletin*, 30, January 25, 1954. In *Makers of Modern Strategy From Machiavelli to the Nuclear Age*. Edited by Peter Paret. Princeton, NJ: Princeton University Press, 1986.

Staudenmaier, William O. "The Decisive Role of Landpower." In *Defense Policy in the Reagan Administration*. Washington, D.C.: National Defense University Press 1988.

Tanter, Raymond. "Chapter Two: Iran: Balance of Power vs. Dual Containment." In *Rogue Regimes: Terrorism and Proliferation*. New York: St. Martins Press, September 1996. On-line. Internet, 12 February 2001. Available from http://www-personal.umich.edu/~rtanter/rogue.iran.html.

Tenet, George J., Director of Central Intelligence. "The Worldwide Threat in 2000: Global Realities on Our National Security." Statement to Senate Select Committee on Intelligence, 2 February 2000. On-line. Internet, 26 February 2001. Available from http://www.usinfo.state.gov/topical/pol/terror/00020201.htm.

Tenet, George, Director of Central Intelligence. Briefing to Senate Select Committee on Intelligence. "Worldwide Threat 2001: National Security in a Changing World," 7 February 2001. On-line. Internet, 5 March 2001. Available from http://www.cia.gov/cia/public_affairs/speeches/UNCLASWWT_02072001.html.

The Heritage Foundation Commission on Missile Defense. "Chapter 3: Fundamentals of Global Defense." In *Defending America: A Plan to Meet the Urgent Missile Threat*. Washington, D.C.: The Heritage Foundation, March 1999. On-line. Internet, 27 February 2001. Available from http://bds.cetin.net.cn:81/cetin2/report/tmd/tmdzl/nmd-US/chapter3.html.

The Heritage Foundation Commission on Missile Defense. "Chapter 4: A Plan for an Affordable and Effective Missile Defense: Recommendations." In *Defending America: A Plan to Meet the Urgent Missile Threat*. Washington, D.C.: The Heritage Foundation, March 1999. On-line. Internet, 27 February 2001. Available from http://bds.cetin.net.cn:81/cetin2/report/tmd/tmdzl/nmd-US/chapter4.html.

The History Place. "The Vietnam War: America Commits 1961-1964." On-line. Internet, 11 February 2001. Available from http://www.historyplace.com/unitedstates/vietnam/index-1961.html.

The President's Commission on Critical Infrastructure Protection. *Fact Sheet: President's Commission on Critical Infrastructure Protection* (1997). On-line. Internet, 25 July 2000. Available from http://www.info-sec.com/pccip/pccip2/backgrd.html.

The President's Commission on Critical Infrastructure Protection. *Our Nation's Critical Infrastructures: Some Working Definitions* (1997). On-line. Internet, 25 July 2000. Available from http://www.info-sec.com/pccip/pccip2/glossary.html.

The White House. *Fact Sheet: Protecting America's Critical Infrastructures: PDD 63* (22 May 1998). On-line. Internet, 25 July 2000. Available from http://www.fas.org/irp/offdocs/pdd-63.htm.

Thompson, Loren. *Homeland Defense: A Confusing Start*, 7 September 1999. On-line. Internet, 30 March 2001. Available from http://www.defensedaily.com/reports/homeland.htm.

Timmerman, Kenneth R. "Is Iran-Saudi Détente Underway?" *Wall Street Journal Europe*, 20 May 1999. On-line. Internet, 26 February 2001. Available from http://www.iran.org/tib/krt/wsje_990520.htm.

Tonello, Alex.. *The Rise and Fall of the Strategic Defense Initiative*. 28 October 1997. On-line. Internet, 11 January 2001. Available from http://members.tripod.com/~atonello/sdi.htm.

Trendafilovski; Vladimir. *A-35 Anti-Ballistic Missile System*. 18 august 1998. On-line. Internet, 13 January 2001. Available from http://www.wonderland.org.nz/a-35.htm.

Trendafilovski; Vladimir. *Russian Anti-Ballistic Guided Missile Systems: RZ-25 Anti-Ballistic Missile System*. 18 August 1998. On-line. Internet, 8 January 2001. Available from http://www.wonderland.org.nz/rz-25.htm.

Trendafilovski; Vladimir. *Russian Anti-Ballistic Guided Missile Systems: SA-5 GRIFFON*. 18 August 1998. On-line. Internet, 8 January 2001. Available from http://www.wonderland.org.nz/rusabgm.htm.

Truman Library, National Archives and Records Administration. *Harry Truman and the Truman Doctrine*. On-line. Internet, 10 February 2001. Available fromhttp://www.trumanlibrary.org/teacher/doctrine.htm.

United States Army National Guard. *The Guard Today – Current Initiatives*, no date. On-line. Internet, 30 March 2001. Available from http://www.arng.ngb.army.mil/Operations/statements/ps/2001/The%20Guard%20Today.htm.

United States Joint Forces Command. *USJFCOM Command Mission*, no date. On-line. Internet, 30 March 2001. Available from http://137.246.33.101/cmdmission2.htm.

U.S. Army Tank-automotive and Armaments Command (TACOM) Security Assistance Center. *The Beginnings of NATO*. On-line. Internet, 22 February 2001. Available from http://www-acala1.ria.army.mil/tsac/nato.htm.

U.S. Army Tank-automotive and Armaments Command (TACOM) Security Assistance Center. *The Eisenhower Doctrine*. On-line. Internet, 10 February 2001. Available from http://www-acala1.ria.army.mil/tsac/eisenhwr.htm.

U.S. Army Tank-automotive and Armaments Command (TACOM) Security Assistance Center. *The Kennedy and Johnson Administrations*. On-line. Internet, 10 February 2001. Available from http://www-acala1.ria.army.mil/tsac/kenjohns.htm.

U.S. Army Tank-automotive and Armaments Command (TACOM) Security Assistance Center. *The Nixon Doctrine*. On-line. Internet, 12 February 2001. Available from http://www-acala1.ria.army.mil/tsac/nixon.htm.

U.S. Army Tank-automotive and Armaments Command (TACOM) Security Assistance Center, *The Carter Administration*. On-line. Internet, 12 February 2001. Available from http://www-acala1.ria.army.mil/tsac/carter.htm.

U.S. Army Tank-automotive and Armaments Command (TACOM) Security Assistance Center. *The Bush Administration*. 21 November 2000. On-line. Internet, 12 February 2001. Available from http://www-acala1.ria.army.mil/tsac/bush.htm.

U.S. Army Training and Doctrine Command (TRADOC). *White Paper: Supporting Homeland Defense*. 18 May 1999. On-line. Internet, 1 March 2001. Available from http://www.fas.org/spp/starwars/program/homeland/final-white-paper.htm.

U.S. Congress. *The Militia Act of 1792*. 2nd Cong., 1st sess., 2 May 1792. On-line. Internet, 13 March 2001. Available from http://www.constitution.org/mil/mil_act_1792.htm.

U.S. Department of State. "Background Information on Terrorist Groups." *Patterns of Global Terrorism 1998*. (Washington, D.C.: Government Printing Office 1999). On-line. Internet, 25 February 2001. Available from http://www.state.gov/www/global/terrorism/1998Report/appb.html.

U.S. Department of State, Office of the Coordinator for Counter-terrorism. *Fact Sheet: Usama bin Laden*. 21 August 1998. On-line. Internet, 16 February 2001. Available from http://www.state.gov/www/regions/africa/fs_bin_ladin.html.

U.S. Department of State, Office of the Coordinator for Counter-terrorism. *Patterns of Global Terrorism 1998*. Washington D.C.: U.S. Government Printing Office, April 1999. On-line. Internet, 23 February 2001, available from http://www.state.gov/www/global/terrorism/1998Report/1998index.html.

U.S. Department of State, Office of the Coordinator for Counter-terrorism. *1996 Patterns of Global Terrorism Report*. Washington, D.C.: U.S. Government Printing Office, 1997. On-line. Internet, 23 February 2001. Available from http://www.state.gov/www/global/terrorism/1996Report/middle.html.

U.S. Department of State, Office of the Coordinator for Counter-terrorism. *Patterns of Global Terrorism 1993*. Washington D.C.: U.S. Government Printing Office 1994. On-line. Internet, 26 February 2001. Available from http://www.hri.org/docs/USSD-Terror/93/statespon.html#Iraq.

U.S. Department of State, Office of the Coordinator for Counter-terrorism. *Patterns of Global Terrorism 1999*. Washington, D.C.: U.S. Government Printing Office, 2000. On-line, Internet, 26 February 2001. Available from http://www.state.gov/www/global/terrorism/1999report/sponsor.html#Iraq.

U.S. Department of State, Office of the Coordinator for Counter-terrorism. *Patterns of Global Terrorism 1999*. Washington, D.C.: U.S. Government Printing Office, April

2000. On-line. Internet, 16 February 2001. Available from http://www.state.gov/www/global/terrorism/1999report/asia.html#Afghanistan.

U.S. Department of State. *Strategic Arms Limitation Talks (SALT I)*. On-line. Internet, 12 January 2001. Available from http://www.state.gov/www/global/arms/treaties/salt1.html.

U.S. Department of State. *Treaty Between The United States Of America And The Union Of Soviet Socialist Republics On The Elimination Of Their Intermediate-Range And Shorter-Range Missiles.* On-line. Internet, 13 March 2001. Available from http://www.state.gov/www/global/arms/treaties/inf1.html.

Walpole, Robert D., National Intelligence Officer for Strategic and Nuclear Programs. "North Korea's Taepo Dong Launch and Some Implications on the Ballistic Missile Threat to the United States." Speech, Center for Strategic and International Studies, Washington D.C., 8 December 1998. On-line. Internet, 26 February 2001. Available from http://www.cia.gov/cia/public_affairs/speeches/archives/1998/walpole_speech_1208 98.html.

Weigley, Russell F. *The American Way of War*. Bloomington, Indiana: Indiana University Press, 1973.

Weinberger, Caspar, Secretary of Defense, and George Schultz. *Soviet Strategic Defense Programs.* October 1985. On-line. Internet, 13 January 2001. Available from http://www.fas.org/irp/dia/product/ssdp.htm.

White, Mary Jo, United States Attorney. "Indictment S(9) 98 Cr. 1023 (LBS) United States of America versus Usama bin Laden [and other associates]." On-line. Internet, 25 February 2001. Available from http://www.fbi.gov/majcases/eastafrica/indictment.pdf.

Williams, John Allen. "The U.S. Navy Under the Reagan Administration." In *Defense Policy in the Reagan Administration*. Washington, D.C.: National Defense University Press 1988.

Williams, John Allen. "The US Navy Under the Reagan Administration and Global Forward Strategy." In *Defense Policy in the Reagan Administration*. Edited by William P. Snyder and James Brown. Washington, D.C.: National Defense University Press 1988.

Woolf, Amy F. "IB98038: Nuclear Weapons in Russia: Safety, Security, and Control Issues." *Congressional Research Service Issue Brief for Congress*, 21 November 2000. On-line. Internet, 28 February 2001. Available from http://www.cnie.org/nle/inter-64.html#_1_3.

York, Herbert. *Civil Defense is Propaganda*. PBS interview. On-line. Internet, 25 September 2000. Available from http://www.pbs.org/wgbh/amex/bomb/filmmore/reference/interview/york2.html.

Young, Rick. "Lessons of Vietnam: A Conversation with Major H.R. McMaster." PBS Online, 1999. On-line. Internet, 11 February 2001. Available from http://www.pbs.org/wgbh/pages/frontline/shows/military/etc/lessons.html.

Zwick, Peter R. "American-Soviet Relations: The Rhetoric and Realism." In *Defense Policy in the Reagan Administration*. Edited by William P. Snyder and James Brown. Washington, D.C.: National Defense University Press 1988.

www.ingramcontent.com/pod-product-compliance
Lightning Source LLC
Chambersburg PA
CBHW080435110426
42743CB00016B/3173